Getting Started with Bluetooth Low Energy

*Kevin Townsend, Carles Cufí,
Akiba, and Robert Davidson*

Beijing · Cambridge · Farnham · Köln · Sebastopol · Tokyo

Getting Started with Bluetooth Low Energy

by Kevin Townsend, Carles Cufí, Akiba, and Robert Davidson

Printed in the United States of America.

Published by O'Reilly Media, Inc., 1005 Gravenstein Highway North, Sebastopol, CA 95472.

O'Reilly books may be purchased for educational, business, or sales promotional use. Online editions are also available for most titles (*http://safaribooksonline.com*). For more information, contact our corporate/institutional sales department: 800-998-9938 or *corporate@oreilly.com*.

Editors: Brian Sawyer and Mike Loukides
Production Editor: Melanie Yarbrough
Proofreader: Eliahu Sussman
Indexer: Judith McConville

Cover Designer: Karen Montgomery
Interior Designer: David Futato
Illustrator: Rebecca Demarest

May 2014:   First Edition

Revision History for the First Edition:

2014-04-29:   First release

2015-04-17:   Second release

See *http://oreilly.com/catalog/errata.csp?isbn=9781491949511* for release details.

ISBN: 978-1-491-94951-1

[LSI]

Table of Contents

Preface

Bluetooth Low Energy (BLE), which was introduced as part of the Bluetooth 4.0 specification, is an exciting wireless technology that gives mobile application developers unprecedented access to external hardware and provides hardware engineers with easy and reliable access to their devices from every major mobile operating system.

This book aims to provide a solid, practical, high-level understanding of Bluetooth Low Energy: how data is organized, how devices communicate with each other, and the key design decisions and tradeoffs that were made by the protocol design teams. It should leave you with enough of an understanding of BLE to approach the high-level APIs on most modern embedded devices and mobile operating systems with confidence and enable you to make sense of the terminology and naming conventions in more in-depth technical documentation when you need to dig deeper. It should also clarify some of the specific strengths and limitations that distinguish BLE from other wireless technologies, such as WiFi, NFC, classic Bluetooth, Zigbee, and so on.

Experienced embedded firmware engineers will leave better prepared to dive deeper into the existing technical documentation, and mobile application developers will have a clearer idea of how data is organized in BLE devices and how to communicate with existing hardware.

Who This Book Is For

This book intends to serve two main audiences:

Mobile application developers
> First, the book serves as a high-level conceptual overview of Bluetooth Low Energy for mobile application developers who want to design applications capable of talking to physical devices in the outside world, but who might not find the official 2,600-page Bluetooth Core Specification 4.1 particularly easy to approach.

Embedded engineers

On the other side of the coin, the book is also for traditional embedded engineers who are considering Bluetooth Low Energy from a product design point of view. If you need to get up to speed quickly on what BLE is and isn't, this book should help you quickly evaluate its strengths and weaknesses as a wireless protocol for your project.

How to Use This Book

This book is organized into three main sections.

Overview of BLE

The first four chapters provide a high-level overview of Bluetooth Low Energy as a technology, explaining how data is organized and what its key limitations are, while also introducing all the key concepts that you're likely to encounter working with BLE:

Chapter 1, Introduction

The first chapter introduces the basic concepts of the wireless standard known as Bluetooth Low Energy. It briefly describes the essentials required for understanding the most important elements of the technology and gives an outline of the different specification and chip configurations that can be found today. This chapter also introduces and explains elementary concepts fundamental to BLE, such as broadcasting, connections, and the different roles that devices can assume.

Chapter 2, Protocol Basics

This chapter focuses on the protocol stack as a whole and the different entities that belong to it. It gives an overview of each of the protocol layers and their essential features, filtering out details from the specification that are not directly relevant to BLE application developers. Each layer is described in the context of the role it assumes as part of the bigger picture, with special attention to the impact it might have in real-life scenarios.

Chapter 3, GAP (Advertising and Connections)

This chapter presents the Generic Access Profile (GAP), which governs the advertising process as well as connections. It gives an overview of the modes and procedures that allow devices to interact using both advertising packets to broadcast information and connections to exchange data.

Chapter 4, GATT (Services and Characteristics)

This chapter provides an overview of the Generic Attribute Profile (GATT), which establishes the hierarchy and format used to represent and manipulate data in BLE. It introduces the fundamental concepts of services and characteristics, as well as the procedures that allow connected devices to exchange data with each other.

Tools for Development and Testing

The next three chapters present useful tools (both hardware and software) for developing or testing BLE-enabled applications or devices. These chapters focus on low-cost, easily accessible tools to help you get started without investing thousands of dollars:

Chapter 5, Hardware Platforms
> This chapter provides product designers with an overview of some of the latest embedded development platforms for BLE peripherals or products.

Chapter 6, Debugging Tools
> Whether you're designing your own device or designing an application that talks to existing hardware, you'll almost certainly have many hours of debugging ahead of you. Debugging wireless devices is a different process than purely software-based development. This chapter presents some useful debugging tools for working with BLE and seeing what's actually being sent over the air.

Chapter 7, Application Design Tools
> This chapter presents key tools for mobile application developers working with BLE. These tools will help you quickly test and verify your software or even simulate devices, if you don't have access to real hardware early in the design process.

Development Platforms

Finally, the last three chapters introduce the main development platforms you are likely to work with for BLE (iOS and Android for application developers, and various embedded platforms for product designers and embedded hardware engineers):

Chapter 8, Android Programming
> This chapter provides a basic overview of the hardware, software, and processes required to implement Bluetooth Low Energy on the Android operating system.

Chapter 9, iOS Programming
> This chapter explores some of the key iOS 7 frameworks, classes, and methods that support BLE application development. Examples explore application development using BLE to read the battery level of a peripheral and an application that uses the iBeacon for location determination.

Chapter 10, Embedded Application Development
> This chapter introduces the tools needed to compile code for embedded devices. Using the nRF51822-EK discussed in Chapter 5 with the free, open source GNU toolchain and cross-compiler for ARM, you'll create a heart rate monitor example to run natively on the nRF51822 SoC.

Conventions Used in This Book

The following typographical conventions are used in this book:

Italic

> Indicates new terms, URLs, email addresses, filenames, and file extensions.

`Constant width`

> Used for program listings, as well as within paragraphs to refer to program elements such as variable or function names, databases, data types, environment variables, statements, and keywords.

`Constant width bold`

> Shows commands or other text that should be typed literally by the user.

`Constant width italic`

> Shows text that should be replaced with user-supplied values or by values determined by context.

 This element signifies a tip, suggestion, or general note.

 This element indicates a warning or caution.

Using Code Examples

Supplemental material (code examples, exercises, etc.) is available for download at *https://github.com/microbuilder/IntroToBLE*.

This book is here to help you get your job done. In general, if example code is offered with this book, you may use it in your programs and documentation. You do not need to contact us for permission unless you're reproducing a significant portion of the code. For example, writing a program that uses several chunks of code from this book does not require permission. Selling or distributing a CD-ROM of examples from O'Reilly books does require permission. Answering a question by citing this book and quoting example code does not require permission. Incorporating a significant amount of example code from this book into your product's documentation does require permission.

We appreciate, but do not require, attribution. An attribution usually includes the title, author, publisher, and ISBN. For example: "*Getting Started with Bluetooth Low Energy* by Kevin Townsend, Carles Cufí, Akiba, and Robert Davidson (O'Reilly). Copyright 2014 Kevin Townsend, Carles Cufí, Akiba, and Robert Davidson, 978-1-491-94951-1."

If you feel your use of code examples falls outside fair use or the permission given above, feel free to contact us at *permissions@oreilly.com*.

Safari® Books Online

 *Safari Books Online* is an on-demand digital library that delivers expert content in both book and video form from the world's leading authors in technology and business.

Technology professionals, software developers, web designers, and business and creative professionals use Safari Books Online as their primary resource for research, problem solving, learning, and certification training.

Safari Books Online offers a range of plans and pricing for enterprise, government, education, and individuals.

Members have access to thousands of books, training videos, and prepublication manuscripts in one fully searchable database from publishers like O'Reilly Media, Prentice Hall Professional, Addison-Wesley Professional, Microsoft Press, Sams, Que, Peachpit Press, Focal Press, Cisco Press, John Wiley & Sons, Syngress, Morgan Kaufmann, IBM Redbooks, Packt, Adobe Press, FT Press, Apress, Manning, New Riders, McGraw-Hill, Jones & Bartlett, Course Technology, and hundreds more. For more information about Safari Books Online, please visit us online.

How to Contact Us

Please address comments and questions concerning this book to the publisher:

O'Reilly Media, Inc.
1005 Gravenstein Highway North
Sebastopol, CA 95472
800-998-9938 (in the United States or Canada)
707-829-0515 (international or local)
707-829-0104 (fax)

We have a web page for this book, where we list errata, examples, and any additional information. You can access this page at *http://bit.ly/gs-with-bluetooth-low-energy*.

To comment or ask technical questions about this book, send email to *bookques tions@oreilly.com*.

For more information about our books, courses, conferences, and news, see our website at *http://www.oreilly.com*.

Find us on Facebook: *http://facebook.com/oreilly*

Follow us on Twitter: *http://twitter.com/oreillymedia*

Watch us on YouTube: *http://www.youtube.com/oreillymedia*

Acknowledgments

Thanks to Clara and Judith, for their endless patience and understanding, to Ha Thach, for his endless help on the CNU codebase for the nRF51822, and to pt and Limor for helping make this possible day after day.

— Kevin Townsend

Thanks to Carla for putting up with the incessant keyboard clicking, and to Vinayak for all I have learned from him over the years.

— Carles Cufí

It's rare I get to work with a great team so I'd like to thank all the coauthors for allowing me to participate in this project and all the O'Reilly staff for giving me a glimpse of all the work that goes into putting a book together. I'd also like to thank the worldwide community of hackers and hackerspaces for providing constant inspiration and the drive to learn more so I can teach workshops. And finally, I'd like to thank the wireless sensor network community and all the crazies that are part of it, many of whom are also part of the hacker community as well as part of this book.

— Akiba

Thanks to my son Joseph and my daughter Leah for giving up the time with me so I could work on this.

— Robert Davidson

CHAPTER 1

Introduction

Bluetooth Low Energy (BLE, also marketed as Bluetooth Smart) started as part of the Bluetooth 4.0 Core Specification. It's tempting to present BLE as a smaller, highly optimized version of its bigger brother, classic Bluetooth, but in reality, BLE has an entirely different lineage and design goals.

Originally designed by Nokia as Wibree before being adopted by the Bluetooth Special Interest Group (SIG), the authors weren't trying to propose another overly broad wireless solution that attempts to solve every possible problem. From the beginning, the focus was to design a radio standard with the lowest possible power consumption, specifically optimized for low cost, low bandwidth, low power, and low complexity.

These design goals are evident through the core specification, which attempts to make BLE a genuine low-power standard, designed to actually be implemented by silicon vendors and used in the real world on a tight energy and silicon budget. It might be the first widely adopted standard that can realistically lay claim to running for an extended period of time off a humble coin cell, though many other wireless technologies regularly make that claim in their marketing.

What Makes BLE Different

While Bluetooth Low Energy is a good technology on its own merit, what makes BLE genuinely exciting—and what has pushed its phenomenal adoption rate so far so quickly —is that it's the right technology, with the right compromises, at the right time. For a relatively young standard (it was introduced in 2010), BLE has seen an uncommonly rapid adoption rate, and the number of product designs that already include BLE puts it well ahead of other wireless technologies at the same point of time in their release cycles.

Compared to other wireless standards, the rapid growth of BLE is relatively easy to explain: BLE has gone further faster because its fate is so intimately tied to the phenomenal growth in smartphones, tablets, and mobile computing. Early and active adoption of BLE by mobile industry heavyweights like Apple and Samsung broke open the doors for wider implementation of BLE.

Apple, in particular, has put significant effort into producing a reliable BLE stack and publishing design guidelines around BLE. This, in turn, pushed silicon vendors to commit their limited resources to the technology they felt was the most likely to succeed or flourish in the long run, and the Apple stamp of approval is clearly a compelling argument when you need to justify every research and development dollar invested.

While the mobile and tablet markets become increasingly mature and costs and margins are decreasing, the need for connectivity with the outside world on these devices has a huge growth potential, and it offers peripheral vendors a unique opportunity to provide innovative solutions to problems people might not even realize that they have today.

So many benefits have converged around BLE, and the doors have been opened wide for small, nimble product designers to gain access to a potentially massive market with task-specific, creative, and innovative products on a relatively modest design budget. You can purchase all-in-one radio-plus-microcontroller (system-on-chip) solutions today for well under $2 per chip and in low volumes, which is well below the total overall price point of similar wireless technologies such as WiFi, GSM, Zigbee, etc. And BLE allows you to design viable products *today* that can talk to any modern mobile platform using chips, tools, and standards that are easy to access.

Perhaps one of the less visible key factors contributing to the success of BLE is that it was designed to serve as an extensible framework to exchange data. This is a fundamental difference with classic Bluetooth, which focused on a strict set of use cases. BLE, on the other hand, was conceived to allow anyone with an idea and a bunch of data points coming from an accessory to realize it without having to know a huge amount about the underlying technology. The smartphone vendors understood the value of this proposition early on, and they provided flexible and relatively low-level APIs to give mobile application developers the freedom to use the BLE framework in any way they see fit.

Devices that talk to smartphones or tablets also offer another easy-to-underestimate advantage for product designers: they have an unusually low barrier to adoption. Users are already accustomed to using the handsets or tablets in their possession, which means the burden of learning a new UI is limited, as long as we respect the rich visual language that people have grown accustomed to in the platforms they use.

With a relatively easy-to-understand data model, no intrusive licensing costs, no fees to access the core specs, and a lean overall protocol stack, it should be clear why platform designers and mobile vendors see a winner in BLE.

The Specification

In June 2010, the Bluetooth SIG introduced Bluetooth Low Energy with version 4.0 of the Bluetooth Core Specfication. The specification had been several years in the making and most of the controversial sections and decisions were finally ironed out by the companies involved in the development process, with a few additional concerns left to be dealt with in subsequent updates of the specification.

The first such major update, Bluetooth 4.1, was released in December 2013 and is the current reference for anyone looking to develop BLE products. Althought the basic building blocks, procedures, and concepts remained intact, this release also introduced multiple changes and improvements to smooth the experience of the user.

As with all Bluetooth specifications, 4.1 is backwards compatible with 4.0, ensuring the correct interoperability among devices implementing different specification versions. The specifications allow developers to release and qualify products against either of the versions (until deprecated), although the rapid adoption of new specification releases and the fact that the 4.1 version standardizes several common practices among devices makes it recommendable to target the latest available one.

Unless otherwise noted, this book uses the Bluetooth 4.1 specification as reference. Wherever necessary, and especially when mentioning a noteworthy change or addition, we will clarify when the previous 4.0 specification does not cover a particular area.

To obtain the latest adopted version of the Bluetooth specification, see the Bluetooth SIG's Specification Adopted Documents page (*http://bit.ly/1gg4UhU*).

Configurations

The Bluetooth specification covers both classic Bluetooth (the well-known wireless standard that has been commonplace in many consumer devices for a number of years now) and Bluetooth Low Energy (the new, highly optimized wireless standard introduced in 4.0). Those two wireless communication standards are *not directly compatible* and Bluetooth devices qualified on any specification version prior to 4.0 cannot communicate in any way with a BLE device. The on-air protocol, the upper protocol layers, and the applications are different and incompatible between the two technologies.

Based on Specification Support

Table 1-1 shows the wireless technologies implemented for the three main device types on the market today.

Table 1-1. Specification configurations

Device	BR/EDR (classic Bluetooth) support	BLE (Bluetooth Low Energy) support
Pre-4.0 Bluetooth	Yes	No
4.x Single-Mode (Bluetooth Smart)	No	Yes
4.x Dual-Mode (Bluetooth Smart Ready)	Yes	Yes

As you can see, the Bluetooth Specification (4.0 and above) defines two wireless technologies:

BR/EDR (classic Bluetooth)
   The wireless standard that has evolved with the Bluetooth Specification since 1.0.

BLE (Bluetooth Low Energy)
   The low-power wireless standard introduced with version 4.0 of the specification.

And these are the two device types that be used with these configurations:

Single-mode (BLE, Bluetooth Smart) device
   A device that implements BLE, which can communicate with single-mode and dual-mode devices, but not with devices supporting BR/EDR only.

Dual-mode (BR/EDR/LE, Bluetooth Smart Ready) device
   A device that implements both BR/EDR and BLE, which can communicate with any Bluetooth device.

Figure 1-1 shows the configuration possibilities between available Bluetooth versions and device types, along with the protocol stacks that allow these devices to communicate with each other.

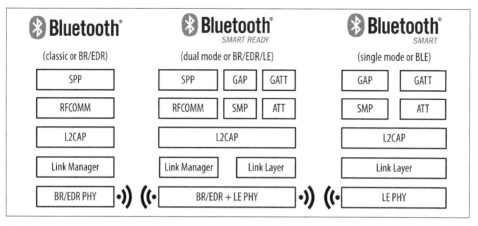

Figure 1-1. Configurations between Bluetooth versions and device types

More and more BR/EDR devices entering the market include BLE as well, and the trend is expected to continue as single-mode BLE sensors become more ubiquitous. Those dual-mode devices can forward the data obtained from a single-mode BLE device to the internet using their GSM or WiFi radios, a feature that is becoming more and more common as more BLE sensors enter the market.

Based on Chip Count

Chapter 2 introduces and discusses the several protocol layers that constitute the Bluetooth protocol stack, but for now it suffices to outline the three main building blocks of every Bluetooth device:

Application
> The user application interfacing with the Bluetooth protocol stack to cover a particular use case.

Host
> The upper layers of the Bluetooth protocol stack.

Controller
> The lower layers of the Bluetooth protocol stack, including the radio.

Additionally, the specification provides a standard communications protocol between the host and the controller—the Host Controller Interface (HCI)—to allow interoperability between hosts and controllers produced by different companies.

These layers can be implemented in a single integrated circuit (IC) or chip, or they can be split in several ICs connected through a communication layer (UART, USB, SPI, or other).

These are the three most common configurations found in commercially available products today:

SoC (system on chip)
> A single IC runs the application, the host, and the controller.

Dual IC over HCI
> One IC runs the application and the host and communicates using HCI with a second IC running the controller. The advantage of this approach is that, since HCI is defined by the Bluetooth specification, any host can be combined with any controller, regardless of the manufacturer.

Dual IC with connectivity device
> One IC runs the application and communicates using a propietary protocol with a second IC running both the host and the controller. Since the specification does not include such a protocol, the application must be adapted to the specific protocol of the chosen vendor.

Figure 1-2 shows the various hardware configurations with the layers of the Bluetooth protocol stack.

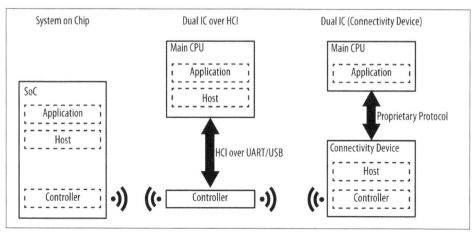

Figure 1-2. Hardware configurations

Simple sensors tend to use SoC configurations to keep the cost and printed circuit board (PCB) complexity low, whereas smartphones and tablets usually opt for the Dual IC over HCI configuration because they usually already have a powerful CPU available to run the protocol stack. The Dual IC with connectivity device configuration is used in other scenarios, one of which could be a watch with a specialized microcontroller to which BLE connectivity is added without overhauling the whole design.

Key Limitations

Like all things in engineering, good design is all about making the right tradeoffs, and Bluetooth Low Energy is no different. BLE doesn't attempt to be a solution to every wireless data transfer need, and classic Bluetooth, WiFi, NFC, and other wireless technologies clearly still have their place, with their own unique set of design tradeoffs and decisions.

To help understand what BLE is (and isn't), it's useful to recognize its key limitations (as defined in the Bluetooth 4.0 specification and later) and how these limitations translate into real-world products.

Data Throughput

The modulation rate of the Bluetooth Low Energy radio is set by the specification at a constant 1Mbps. This sets the *theoretical* upper limit for the throughput that BLE can provide, but in actual terms, this limit is typically lowered significally by a variety of factors, including but not restricted to bidirectional traffic, protocol overhead, CPU and radio limitations, and artificial software restrictions.

To illustrate some of these practical restrictions, consider the following basic preconditions we'll use for a calculation:

- A central (master) device has initiated and established a connection with a peripheral (slave) accessory.
- While in an active connection, the specification defines the connection interval to be the interval between two consecutive connection events (a data exchange before going back to an idle state to save power), and this connection interval can be set to a value between 7.5 ms and 4 s.

"Link Layer" on page 17 and "Roles" on page 36 discuss the different roles within a connection in detail. For this example, we'll use the nRF51822, a widely available SoC (system on chip) BLE IC manufactured by Nordic Semiconductor that is used in a variety of BLE accessories on the market. Nordic's radio hardware and BLE stack impose the following data throughput limitations:

- The nRF51822 can transmit up to six data packets per connection interval (limited by the IC).
- Each outgoing data packet can contain up to 20 bytes of user data (set by the specification unless higher packet sizes are negotiated).

Assuming the shortest connection interval (the frequency at which the master and the slave exchange packets, described in "Connections" on page 22) of 7.5 ms, this provides a maximum of 133 connection events (a single packet exchange between the two peers) per second and 120 bytes per connection event (6 packets * 20 user bytes per packet). Continuously transmitting at the maximum data rate of the nRF51822 would suggest the following real-world calculation:

```
133 connection events per second * 120 bytes = 15960 bytes/s
                                    or ~0.125Mbit/s (~125kbit/s)
```

That's already significantly lower than the theoretical maximum of BLE, but the peer device you are pushing data to (typically a smart device such as a smartphone or a tablet) can add further limitations.

Your smartphone or tablet might also be busy talking to other devices, and vendor-implemented BLE stacks inevitably have their own limitations, which means the central device might not actually be able to handle data at the maximum data rate either. And because of multiple other factors, the actual connection interval might be spread out further or more irregularly than you had originally planned.

So, in practice, a typical best-case scenario should probably assume a potential maximum data throughpout in the neighborhood of 5-10 KB per second, depending on the limitations of both peers. This should give you an idea of what you can and can't do with Bluetooth Low Energy in terms of pushing data out to your phone or tablet and explain why other technologies such as WiFi and classic Bluetooth still have their place in the world.

Racing to Idle

In a world where things usually get faster with time, 10 KB/s might seem slow and counterproductive, but it does highlight the primary design goal of Bluetooth Low Energy: *low energy*! Even transmitting at these relatively modest data rates, 10 KB/s will quickly drain any small coin cell battery, and the Bluetooth SIG made a clear, conscious effort not to design yet another generic wireless protocol and slap the label *low power* on it. Instead, they wanted to design the lowest power protocol possible, optimizing in every way possible to acheive that goal. The easiest way to avoid consuming precious battery power is to turn the radio off as often as possible and for as long as possible, and that is achieved by using short burst of packets (during a connection event) at a certain frequency (determined by the connection interval). In between those, the radio is simply powered off.

This means low amounts of data transmitted in short bursts, with connection intervals spread as far apart as possible to save battery life. The user-selectable 7.5 ms–4 s connection interval offers a sufficiently wide window to allow product designers to make the right tradeoff between responsiveness (a short connection interval) and battery life (a longer connection interval), without straying to far from the narrow design goals behind BLE.

Operating Range

The actual range of any wireless device depends on a wide variety of factors (operating environment, antenna design, enclosure, device orientation, etc.) but Bluetooth Low Energy is unsurprisingly focused on very short-range communication.

Transmit power (typically measured in dBm) is usually configurable over a certain range (usually between -30 and 0 dBm), but the higher the transmit power (better range), the more demands are placed on the battery, reducing the usable lifetime of the battery cell(s).

It's possible to create and configure a BLE device that can reliably transmit data 30 meters or more *line-of-sight*, but a typical operating range is probably closer to 2 to 5 meters, with a conscious effort to reduce the range and save battery life without the transmission distance becoming a nuisance to the end user.

Network Topology

A Bluetooth Low Energy device can communicate with the outside world in two ways: *broadcasting* or *connections*. Each mechanism has its own advantages and limitations, and they are both subject to the guidelines established by the Generic Access Profile (GAP), which Chapter 3 describes in detail.

Broadcasting and Observing

Using connectionless *broadcasting*, you can send data out to any scanning device or receiver in listening range. As illustrated in Figure 1-3, this mechanism essentially allows you to send data out *one-way* to anyone or anything that is capable of picking up the transmitted data.

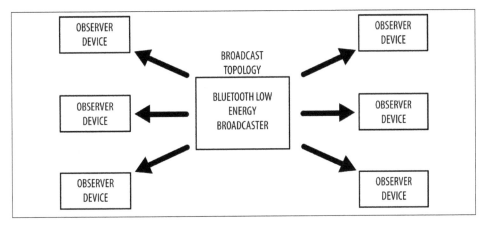

Figure 1-3. Broadcast topology

Broadcasting defines two separate roles:

Broacaster

Sends nonconnectable *advertising* packets periodically to anyone willing to receive them.

Observer

Repeatedly scans the preset frequencies to receive any nonconnectable advertising packets currently being broadcasted.

Broadcasting is important to understand, because it's the only way for a device to transmit data to more than one peer at time. You broadcast data out by taking advantage of the the advertising features of BLE, as discussed in more detail in "Advertising and Scanning" on page 19 and "Broadcast and Observation" on page 38.

The standard advertising packet contains a 31-byte payload used to include data that describes the broadcaster and its capabilities, but it can also include any custom information you want to broadcast to other devices. If this standard 31-byte payload isn't large enough to fit all of the required data, BLE also supports an optional secondary advertising payload (called the *Scan Response*), which allows devices that detect a broadcasting device to request a second advertising frame with another 31-byte payload, for up to 62 bytes total.

Broadcasting is fast and easy to use, and it's a good choice if you want to push only a small amount of data on a fixed schedule or to multiple devices. Chapter 9 provides a practical example of BLE's connectionless broadcasting in action with iBeacon.

A major limitation of broadcasting, when compared to a regular connection, is that there are no security or privacy provisions at all with it (any observer device is able to receive the data being broadcasted), so it might not be suited for sensitive data.

Connections

If you need to transmit data in both directions, or if you have more data than the two advertising payloads can accomodate, you will need to use a *connection*. A *connection* is a permanent, periodical data exchange of packets between *two* devices. It is therefore inherently private (the data is sent to and received by only the two peers involved in a connection, and no other device unless it's indiscriminately sniffing). "Connections" on page 22 provides more information about connections at the lower level, and "Roles" on page 36 discusses the corresponding GAP roles.

Connections involve two separate roles:

Central (master)
> Repeatedly scans the preset frequencies for connectable advertising packets and, when suitable, initates a connection. Once the connection is established, the central manages the timing and initiates the periodical data exchanges.

Peripheral (slave)
> A device that sends connectable advertising packets periodically and accepts incoming connections. Once in an active connection, the peripheral follows the central's timing and exchanges data regularly with it.

To initiate a connection, a central device picks up the connectable advertising packets from a peripheral and then sends a request to the peripheral to establish an exclusive connection between the two devices. Once the connection is established, the peripheral stops advertising and the two devices can begin exchanging data in both directions, as shown in Figure 1-4.

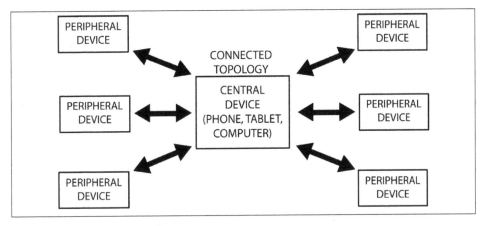

Figure 1-4. Connected topology

A connection is therefore nothing more than the periodical exchange of data at certain specific points in time (connection events) between the two peers involved in it. It is important to note that, although the central is the device that manages the connection establishment, data can be sent independently by either device during each connection event, and the roles do not impose restrictions in data throughput or priority.

Beginning with version 4.1 of the specification, any restrictions on role combinations have been removed, and the following are all possible:

- A device can act as a central and a peripheral at the same time.
- A central can be connected to multiple peripherals.
- A peripheral can be connected to multiple centrals.

Previous versions of the specification limited the peripheral to a single central connection (although not conversely) and limited the role combinations.

The biggest advantage of connections (when compared to broadcasting) is the ability to organize the data with much finer-grained control over each field or property through the use of additional protocol layers and, more specifically, the Generic Attribute Profile (GATT). Data is organized around units called *services* and *characteristics* (discussed in more detail in Chapter 4).

The key thing to keep in mind is that you can have multiple services and characteristics, organized in a meaningful structure. Services can contain multiple characteristics, each with their own access rights and descriptive metadata. Additional advantages include higher throughput, the ability to establish a secure encrypted link, and negotiation of connection parameters to fit the data model.

Connections allow for a much richer, layered data model. They also have the potential to use much less power than broadcast mode because they can extend the delay between connection events further out, or push large chunks of data out only when new values are available, rather than having to continually advertise the full payload at a specific rate without knowing who is listening or how often. Not only that, but the fact that both peers know when the connection events are going to take place in the future allows the radio to be turned off for longer, potentially saving battery power when compared to broadcasting.

Finally, these topologies can be mixed freely in a wider BLE network, as shown in Figure 1-5. A BR/EDR/LE-capable device can bridge together BLE and BR/EDR connections, and the number of combinations and participants on the network is constrained only by the limitations of the radios and protocol stacks of each device taking part in it.

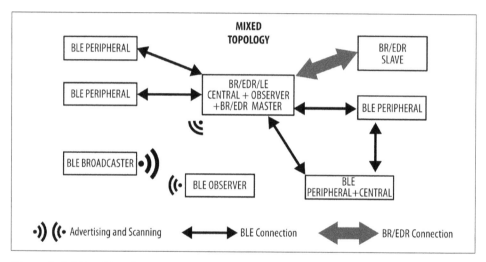

Figure 1-5. Mixed topology

More advanced dual-mode and single-mode devices are starting to appear, devices that are able to combine multiple roles concurrently. This allows them to participate in several connections at a time, while also using advertising to broadcast data.

Protocols versus Profiles

From its inception, the Bluetooth specification introduced a clear separation between the distinct concepts of *protocols* and *profiles*:

Protocols

Building blocks used by all devices conformant to the Bluetooth specification, protocols are the layers that implement the different packet formats, routing, multiplexing, encoding, and decoding that allow data to be sent effectively between peers.

Profiles

"Vertical slices" of functionality covering either basic modes of operation required by all devices (Generic Access Profile, Generic Attribute Profile) or specific use cases (Proximity Profile, Glucose Profile), profiles essentially define how protocols should be used to achieve a particular goal, whether generic or specific.

Chapter 2 covers protocols in detail, but the following sections provide a quick introduction to profiles and what they mean for an application developer.

Generic Profiles

Generic profiles are defined by the specification, and it's important to understand how two of them are fundamental to ensuring interoperability between BLE devices from different vendors:

Generic Access Profile (GAP)

Covering the usage model of the lower-level radio protocols to define roles, procedures, and modes that allow devices to broadcast data, discover devices, establish connections, manage connections, and negotiate security levels, GAP is, in essence, the topmost control layer of BLE. This profile is mandatory for all BLE devices, and all must comply with it.

Generic Attribute Profile (GATT)

Dealing with data exchange in BLE, GATT defines a basic data model and procedures to allow devices to discover, read, write, and push data elements between them. It is, in essence, the topmost data layer of BLE.

GAP (discussed in more detail in Chapter 3) and GATT (discussed in more detail in Chapter 4) are so fundamental to BLE that they are often used as the foundation for application programmer interfaces (APIs) as the entry point for the application to interact with the protocol stack.

Use-Case-Specific Profiles

Use-case-specific profiles in this section and elsewhere are limited to GATT-based profiles. This means that all of these profiles use the procedures and operating models of the GATT profile as a base building block for all further extensions.

No non-GATT profiles exist at the time of this writing, but the introduction of L2CAP connection-oriented channels in version 4.1 of the specification might mean that GATT-less profiles might start to appear in the future.

SIG-defined GATT-based profiles

The Bluetooth SIG goes beyond providing a solid reference framework for the topmost control and data layers of devices involved in a BLE network. Just like the USB specification, it also provides a predefined set of use-case profiles, based on GATT, that completely cover all procedures and data formats required to implement a wide range of specific use cases, including the following:

Find Me Profile
Allows devices to physically locate other devices (use a keyring to find the phone or vice versa).

Proximity Profile
Detects the presence or absence of nearby devices (beep if an item is forgotten when leaving a room).

HID over GATT Profile
Transfers HID data over BLE (keyboards, mice, remote controls).

Glucose Profile
Securely transfers glucose levels over BLE.

Health Thermometer Profile
Transfers body temperature readings over BLE.

Cycling Speed and Cadence Profile
Allows sensors on a bicycle to transfer speed and cadence data to a smartphone or tablet.

A full list of SIG-approved profiles is available at the Bluetooth SIG's Specification Adopted Documents page (*https://www.bluetooth.org/en-us/specification/adopted-specifications*). Additionally, you can browse Bluetooth services and characteristics directly at the Bluetooth Developer Portal (*https://developer.bluetooth.org*) and, more specifically, the list of all currently adopted services (*https://developer.bluetooth.org/gatt/services/Pages/ServicesHome.aspx*).

Vendor-Specific Profiles

The Bluetooth specification also allows vendors to define their own profiles for use cases not covered by the SIG-defined profiles. Those profiles can be kept private to the two peers involved in a particular use case (for example, a health accessory and a smartphone application), or they can also be published by the vendor so that other parties can provide implementations of the profile based on the vendor-supplied specification.

Some examples of the latter include Apple's iBeacon (see "iBeacon" on page 132 for more detail) and Apple Notification Center Service (see "Apple Notification Center Service with an External Display" on page 138).

Protocol Basics

Although users will usually interface directly only with the upper layers of the Bluetooth Low Energy protocol stack, it's probably best to begin with a basic overview of the complete stack, which provides a solid foundation to understanding how and why things operate the way they do.

As shown in Figure 2-1, a complete single-mode BLE device is divided into three parts: controller, host, and application.

Each of these basic building blocks of the protocol stack is split into several layers that provide the functionality required to operate:

Application
> The application, like in all other types of systems, is the highest layer and the one responsible for containing the logic, user interface, and data handling of everything related to the actual use-case that the application implements. The architecture of an application is highly dependent on each particular implementation.

Host
> Includes the following layers:
>
> - Generic Access Profile (GAP)
> - Generic Attribute Profile (GATT)
> - Logical Link Control and Adaptation Protocol (L2CAP)
> - Attribute Protocol (ATT)
> - Security Manager (SM)
> - Host Controller Interface (HCI), Host side

Controller
> Includes the following layers:

- Host Controller Interface (HCI), Controller side
- Link Layer (LL)
- Physical Layer (PHY)

In this chapter, the order of sections will describe the different parts that make up a BLE device from bottom (antenna) to top (user interface).

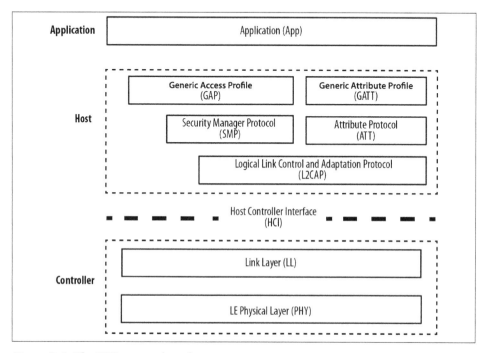

Figure 2-1. The BLE protocol stack

Physical Layer

The physical (PHY) layer is the part that actually contains the analog communications circuitry, capable of modulating and demodulating analog signals and transforming them into digital symbols.

The radio uses the 2.4 GHz ISM (Industrial, Scientific, and Medical) band to communicate and divides this band into 40 channels from 2.4000 GHz to 2.4835 GHz. As shown in Figure 2-2, 37 of these channels are used for connection data and the last three channels (37, 38, and 39) are used as advertising channels to set up connections and send broadcast data.

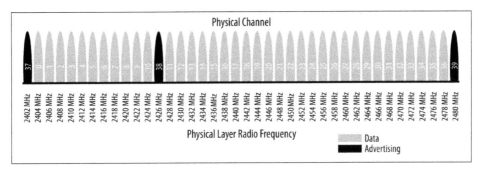

Figure 2-2. Frequency channels

The standard uses a technique called *frequency hopping spread spectrum*, in which the radio *hops* between channels on each connection event using the following formula:

channel = (curr_channel + hop) mod 37

The value of the *hop* is communicated when the connection is established and is therefore different for every new established connection. This technique minimizes the effect of any radio interference potentially present in the 2.4 GHz band across any single channel, especially since WiFi and classic Bluetooth are prevalent in this band and devices might experience heavy interference near devices with a strong transmission power.

The modulation chosen to encode the bitstream over the air is Gaussian Frequency Shift Keying (GFSK), the same modulation used by classic Bluetooth and several other proprietary low-power wireless protocols. The modulation rate for Bluetooth Low Energy is fixed at 1 Mbit/s, which is therefore the upper physical throughput limit for the technology.

In practice, however, as with any other protocol stack, this upper limit is never actually reached when it comes to application throughput, due mainly to protocol overheads in each of the different layers.

Link Layer

The Link Layer is the part that directly interfaces with the PHY, and it is usually implemented as a combination of custom hardware and software. It is also the only hard real-time constrained layer of the whole protocol stack, since it is responsible for complying with all of the timing requirements defined by the specification. It is therefore usually kept isolated from the higher layers of the protocol stack by means of a standard interface that hides the complexity and real-time requirements from the rest of the layers (see "Host Controller Interface (HCI)" on page 24).

Computationally expensive, easily automated functionality is typically implemented in hardware by silicon vendors to avoid overloading the central processing unit that runs all of the software layers in the stack. This functionality usually includes:

- Preamble, Access Address, and air protocol framing
- CRC generation and verification
- Data whitening
- Random number generation
- AES encryption

The software half of the Link Layer manages the link state of the radio, which is how the device connects to other devices. A BLE device can be a master, a slave, or both, depending on the use case and requirements. Devices that initiate connections will be masters and devices that advertise their availability and accept connections will be slaves.

A master can connect to multiple slaves and a slave can be connected to multiple masters. Typically, devices such as smartphones or tablets tend to act as a master, while smaller, simpler, and memory-constrained devices such as standalone sensors usually adopt the slave role.

Bluetooth Low Energy has an inherent asymmetry in its lower layers between master and slave devices, because it requires more resources to act as a master. This asymmetry is similar to USB, in which USB hosts require more resources than USB devices. This type of architectural asymmetry allows low-cost peripherals running on cheap micro-controllers and radios, while the majority of the low-level protocol complexity occurs on devices with more resources, such as smartphones and tablets.

The Link Layer defines the following roles:

Advertiser
    A device sending advertising packets.

Scanner
    A device scanning for advertising packets.

Master
    A device that initiates a connection and manages it later.

Slave
    A device that accepts a connection request and follows the master's timing.

These roles can be logically grouped into two pairs: advertiser and scanner (when not in an active connection) and master and slave (when in a connection).

Bluetooth Device Address

The fundamental identifier of a Bluetooth device, similar to an Ethernet Media Access Control (MAC) adddress, is the *Bluetooth device address*. This 48-bit (6-byte) number uniquely identifies a device among peers. There are two types of device addresses, and one or both can be set on a particular device:

Public device address
> This is the equivalent to a fixed, BR/EDR, factory-programmed device address. It must be registered with the IEEE Registration Authority and will never change during the lifetime of the device.

Random device address
> This address can either be preprogrammed on the device or dynamically generated at runtime. It has many practical uses in BLE, as discussed in more detail in "Address Types" on page 44.

Each procedure must be performed using one of the two, to be specified by the host.

Advertising and Scanning

BLE has only one packet format and two types of packets (*advertising* and *data* packets), which simplifies the protocol stack implementation immensely. Advertising packets serve two purposes:

- To broadcast data for applications that do not need the overhead of a full connection establishment
- To discover slaves and to connect to them

Each advertising packet can carry up to 31 bytes of advertising data payload, along with the basic header information (including Bluetooth device address). Such packets are simply broadcast blindly over the air by the advertiser without the previous knowledge of the presence of any scanning device. They are sent at a fixed rate defined by the advertising interval, which ranges from 20 ms to 10.24 s. The shorter the interval, the higher the frequency at which advertising packets are broadcast, leading to a higher probability of those packets being received by a scanner, but higher amounts of packets transmitted also translate to higher power consumption.

Because advertising uses a maximum of three frequency channels and the advertiser and the scanner are not synchronized in any way, an advertising packet will be received successfully by the scanner only when they randomly overlap, as shown in Figure 2-3.

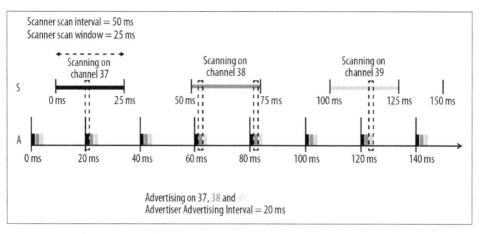

Figure 2-3. Advertising and scanning

The scan interval and scan window parameters define how often and for how long a scanner device will listen for potential advertising packets. As with the advertising interval, those values have a deep impact on power consumption, since they directly relate to the amount of time the radio must be turned on.

The specification defines two basic types of scanning procedures:

Passive scanning

> The scanner simply listens for advertising packets, and the advertiser is never aware of the fact that one or more packets were actually received by a scanner.

Active scanning

> The scanner issues a Scan Request packet after receiving an advertising packet. The advertiser receives it and responds with a Scan Response packet. This additional packet doubles the effective payload that the advertiser is able to send to the scanner, but it is important to note that this does *not* provide a means for the scanner to send any user data at all to the advertiser.

Figure 2-4 illustrates the difference between passive and active scanning.

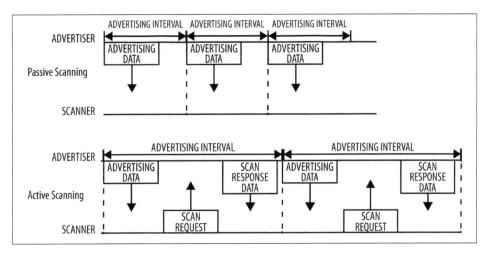

Figure 2-4. Active and passive scanning

Advertising packet types can be classified according to three different properties. The first is *connectability*:

Connectable

A scanner can initate a connection upon reception of such an advertising packet.

Non-connectable

A scanner cannot initiate a connection (this packet is intented for broadcast only).

The second property is *scannability*:

Scannable

A scanner can issue a scan request upon reception of such an advertising packet.

Non-scannable

A scanner cannot issue a scan request upon reception of such an advertising packet.

And the third is *directability*:

Directed

A packet of this type contains only the advertiser's and the target scanner's Bluetooth Addresses in its payload. No user data is allowed. All directed advertising packets are therefore connectable.

Undirected

A packet of this type is not targeted at any particular scanner, and it can contain user data in its payload.

Table 2-1 shows the different advertising packet types and their properties.

Table 2-1. Avertising Packet Types

Advertising Packet Type	Connectable	Scannable	Directed	GAP Name
ADV_IND	Yes	Yes	No	Connectable Undirected Advertising
ADV_DIRECT_IND	Yes	No	Yes	Connectable Directed Advertising
ADV_NONCONN_IND	No	No	No	Non-connectable Undirected Advertising
ADV_SCAN_IND	No	Yes	No	Scannable Undirected Advertising

The advertising packet types are used by the upper layers and, more specifically, GAP to differentiate between operating modes and to define procedures. Therefore, "Modes and Procedures" on page 37 makes extensive use of them at its core.

Connections

To establish a connection, a master first starts scanning to look for advertisers that are currently accepting connection requests. The advertising packets can be filtered by Bluetooth Address or based in the advertising data itself. When a suitable advertising slave is detected, the master sends a connection request packet to the slave and, provided the slave responds, establishes a connection. The connection request packet includes the frequency hop increment, which determines the hopping sequence that both the master and the slave will follow during the lifetime of the connection.

A *connection* is simply a sequence of data exchanges between the slave and the master at predefined times. Shown in Figure 2-5, each exchange is called a *connection event*.

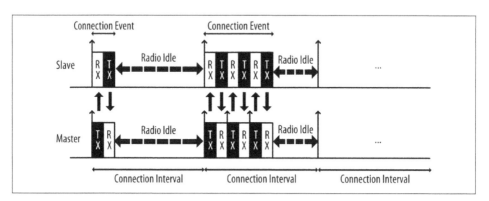

Figure 2-5. Connection events

The following three *connection parameters* are another set of key variables communicated by the master during the establishment of a connection:

Connection interval

The time between the beginning of two consecutive connection events. This value ranges from 7.5 ms (high throughput) to 4 s (lowest possible throughput but also least power hungry).

Slave latency

The number of connection events that a slave can choose to skip without risking a disconnection.

Connection supervision timeout

The maximum time between two received valid data packets before a connection is considered lost.

Because many BLE devices might exist in a given area, or even just for security reasons (in which the master or the slave might be interested in only a small set of preknown devices), the Link Layer implements a *white list* feature, which specifies device addresses of interest to the advertiser or the scanner. Any advertising (if a scanner) or connection request (if an advertiser) packets received from devices whose Bluetooth Address is not present in the white list will simply be dropped.

White Lists

An important feature available in BLE controllers, *white lists* allow hosts to filter devices when advertising, scanning, and establishing connections on both sides. White lists are simply arrays of Bluetooth device addresses that are populated by the host and stored and used in the controller.

A device scanning or initiating a connection can use a white list to limit the number of devices that will be detected or with which it can connect, and the advertising device can use a white list to specify which peers it will accept an incoming connection from. The setting that defines whether a white list is to be used or not is called a *filter policy*. This essentially acts as a switch to turn white list filtering on and off.

Data packets are the workhorse of the protocol and are used to transport user data bidirectionally between the master and slave. These packets have a usable data payload of 27 bytes, but additional procotols further up the stack typically limit the actual amount of user data to 20 bytes per packet, although that logically depends on the protocol being used.

It is important to note that the Link Layer acts as a reliable data bearer. All packets received are checked against a 24-bit CRC and retransmissions are requested when the error checking detects a transmission failure. There is no upper limit for retransmissions; the Link Layer will resend the packet until it is finally acknowledged by the receiver.

Other than advertising, scanning, establishing (and tearing down) connections, and transmitting and receiving data, the Link Layer is also responsible for several control procedures, including these two critical processes:

Changing the connection parameters

Each connection is established with a given set of connection parameters set by the master, but conditions and requirements might change during the lifetime of the connection. A slave might suddenly require a higher throughput for a short burst of data, or conversely, it might detect that in the near future a longer connection interval will suffice to keep the connection alive. The Link Layer allows the master and the slave to request new connection parameters and, in the case of the master, to set them unilaterally at any time. That way, each connection can be fine-tuned to provide the best balance between throughput and power consumption.

Encryption

Security is critical in BLE, and the Link Layer provides the means to exchange data securely over an encrypted link. The keys are generated and managed by the host, but the Link Layer performs the actual data encryption and decryption transparently to the upper layers.

These two procedures are especially relevant, because they each require involvement from the host on both sides to be carried out. The Link Layer handles additional procedures to exchange version information and capabilities internally, so they are transparent to both the host and application developer.

Host Controller Interface (HCI)

As described in Chapter 1, the Bluetooth specification allows several possible configurations based on chip count, and the Host Controller Interface (HCI) is a standard protocol that allows for the communication between a host and a controller to take place across a serial interface. It makes sense to draw a line at that level because, as mentioned earlier in this chapter, the controller is the only module with hard real-time requirements and contact with the physical layer. This means that it is often practical to separate it from the host, which implements a more complex but less timing-stringent protocol stack better suited for more advanced CPUs.

Typical examples of this configuration include most smartphones, tablets, and personal computers, in which the host (and the application) runs in the main CPU, while the controller is located in a separate hardware chip connected via a UART or USB. This is similar to the model used by other technologies, such as WiFi or Ethernet: the TCP/IP stack runs on the main processor, while the lower-level layers execute in a separate IC.

The Bluetooth specification defines HCI as a set of commands and events for the host and the controller to interact with each other, along with a data packet format and a set of rules for flow control and other procedures. Additionally, the spec defines several *transports*, each of which augments the HCI protocol for a specific physical transport (UART, USB, SDIO, etc.).

Semiconductor technology has become inexpensive enough to allow single chips to incorporate the complete controller, host, and application in a single package (a *system-on-chip*, or SoC). In many embedded device applications, heavy integration is preferable, to reduce cost and size on the final device. In the case of BLE, it is common to implement the sensor using a single chip that runs all three layers concurrently on a low-power CPU.

Logical Link Control and Adaptation Protocol (L2CAP)

The rather cryptically named Logical Link Control and Adaptation Protocol (L2CAP) provides two main pieces of functionality. First, it serves as a protocol multiplexer that takes multiple protocols from the upper layers and encapsulates them into the standard BLE packet format (and vice versa).

It also performs fragmentation and recombination, a process by which it takes large packets from the upper layers and breaks them up into chunks that fit into the 27-byte maximum payload size of the BLE packets on the transmit side. On the reception path, it receives multiple packets that have been fragmented and recombines them into a single large packet that will then be sent upstream to the appropriate entity in the upper layers of the host. To draw a simple comparison, L2CAP is similar to TCP, in that it allows a wide range of protocols to seamlessly coexist through a single physical link, each with a different packet size and requirements.

For Bluetooth Low Energy, the L2CAP layer is in charge of routing two main protocols: the Attribute Protocol (ATT) and the Security Manager Protocol (SMP). The ATT (discussed in "Attribute Protocol (ATT)") forms the basis of data exchange in BLE applications, while the SMP (see "Security Manager (SM)" on page 28) provides a framework to generate and distribute security keys between peers.

In addition to those, and since version 4.1 of the specification, L2CAP can create its own user-defined channels for high-throughput data transfer that do not require the additional complexity added by ATT. Initially designed for file transfer, this feature is known as LE Credit Based Flow Control Mode and opens up the possibility of establishing low-latency, high-volume data channels over a BLE connection for applications that require it.

From an application developer's point of view, it is important to note that, whenever only default packet sizes are used, the L2CAP packet header takes up four bytes, which means that the effective user payload length is 27 - 4 = 23 bytes (where 27 bytes is the Link Layer's payload size, as described in "Connections" on page 22).

Attribute Protocol (ATT)

The Attribute Protocol (ATT) is a simple client/server stateless protocol based on *attributes* presented by a device. In BLE, each device is a client, a server, or both, irrespective of whether it's a master or slave. A client requests data from a server, and a server sends data to clients. The protocol is strict when it comes to its sequencing: if a request is still pending (no response for it has been yet received) no further requests can be sent until the response is received and processed. This applies to both directions independently in the case where two peers are acting both as a client and server.

Each server contains data organized in the form of attributes, each of which is assigned a 16-bit attribute handle, a universally unique identifier (UUID), a set of permissions, and finally, of course, a value. The *attribute handle* is simply an identifier used to access an attribute value. The UUID specifies the type and nature of the data contained in the value. For more information, see "UUIDs" on page 52 and "Attributes" on page 53.

When a client wants to read or write attribute values from or to a server, it issues a read or write request to the server with the handle. The server will respond with the attribute value or an acknowledgement. In the case of a read operation, it is up to the client to parse the value and understand the data type based on the UUID of the attribute. On the other hand, during a write operation, the client is expected to provide data that is consistent with the attribute type and the server is free to reject the operation if that is not the case.

ATT operations

The set of operations possible over ATT fall within the following categories:

Error Handling
> Used by the server to respond to any of the requests when an error occurs, this includes only:

> *Error Response*
>> Sent as a response to a request in lieu of the corresponding operation response whenever an error prevented the request from being executed on the server.

Server Configuration
> Used to configure the ATT protocol itself, this includes only:

> *Exchange MTU Request/Response*
>> Exchange between client and server of their respective Maximum Transmission Units (MTU or maximum packet size accepted).

Find Information
> Used by the client to obtain information about the layout of the server's attributes, they include:

Find Information Request/Response
Obtain a list of all attributes in a particular handle range.

Find by Type Value
Obtain the handle range between an attribute identifed by its UUID and its value and the next group delimiter.

Read Operations
Used by the client to obtain the value of one or more attributes, they include:

Read by Type Request/Response
Obtain the value of one or more attributes using a UUID.

Read Request/Response
Obtain the value of attributes using a handle.

Read Blob Request/Response
Obtain part of a value of a long attribute using a handle.

Read Multiple Request/Response
Obtain the value of one or more attributes using multiple handles.

Read by Group Type Request/Response
Similar to Read by Type, but the UUID must be of a grouping type.

Write Operations
Used by the client to set the value of one or more attributes, they include:

Write Request/Response
Write to the value of an attribute and expect a response from the server.

Write Command
Write to the value of an attribute without any response or acknowledgement. This operation does not follow the request/response sequencing and can be sent at any time.

Signed Write Command
Similar to Write Command, but using a signature as described in "Security Manager (SM)" on page 28. This operation does not follow the request/response sequencing and can be sent at any time.

Queued Writes
Used by the client to write to attribute values that are longer than what can fit in a single packet, they include:

Prepare Write Request/Response
Queue a write operation in the server for a particular handle, after which the successful queuing is acknowledged by the server.

Execute Write Request/Response
> Execute all pending queued write operations, the server then reports the success or failure to the client.

Server Initiated
> Used by the server to asynchronously *push* attribute values to the client, they include:

Handle Value Indication/Confirmation
> Asynchronous server update of an attribute's value and identified by its handle, expects an acknowledgement in the form of a confirmation from the client.

Handle Value Notification
> Asynchronous server update of an attribute's value and identified by its handle, without acknowledgment. This operation does not follow the request/response sequencing and can be sent at any time.

All operations except the ones in the *server initiated* category (and a few select others) are grouped into request/response pairs. Requests are always sent by the client and responses are issued by the server as a reply to a request.

Chapter 4 provides more information on ATT itself and how it enables the Generic Attribute Profile (GATT).

Security Manager (SM)

The Security Manager (SM) is both a protocol and a series of security algorithms designed to provide the Bluetooth protocol stack with the ability to generate and exchange security keys, which then allow the peers to communicate securely over an encrypted link, to trust the identity of the remote device, and finally, to hide the public Bluetooth Address if required to avoid malicious peers tracking a particular device.

The Security Manager defines two roles:

Initiator
> Always corresponds to the Link Layer master and therefore the GAP central.

Responder
> Always corresponds to the Link Layer slave and therefore the GAP peripheral.

Although it is always up to the initiator to trigger the beginning of a procedure, the responder can asynchronously request the start of any of the procedures listed in "Security Procedures". There are no guarantees for the responder that the initiator will actually heed the request, serving more as a hint than a real, binding request. This *security request* can logically be issued only by the slave or peripheral end of the connection.

Security Procedures

The Security Manager provides support for the following three procedures:

Pairing

> The procedure by which a *temporary* common security encryption key is generated to be able to switch to a secure, encrypted link. This temporary key is not stored and is therefore not reusable in subsequent connections.

Bonding

> A sequence of *pairing* followed by the generation and exchange of permanent security keys, destined to be stored in nonvolatile memory and therefore creating a permanent bond between two devices, which will allow them to quickly set up a secure link in subsequent connections without having to perform a bonding procedure again.

Encryption Re-establishment

> After a bonding procedure is complete, keys might have been stored on both sides of the connection. If encryption keys have been stored, this procedure defines how to use those keys in subsequent connections to re-establish a secure, encrypted connection without having to go through the pairing (or bonding) procedure again.

Pairing can therefore create a secure link that will last only for the lifetime of the connection, whereas bonding actually creates a permanent association (also called *bond*) in the form of shared security keys that will be used in later connections until either side decides to delete them. Certain APIs and their documentation sometimes use the term *pairing with bonding* instead of simply *bonding*, because a bonding procedure always includes a pairing phase first.

Figure 2-6 shows the two phases of a pairing procedure and the additional phase required for a bonding procedure.

Initially (Phase 1), all information required to generate the temporary key is exchanged between the two devices. Next, (Phase 2) the actual temporary encryption key (Short Term Key or STK) is generated on both sides independently and then used to encrypt the connection. Once the connection is secured by encryption, and only if performing bonding, the permanent keys can be distributed for storage and reuse at a later time.

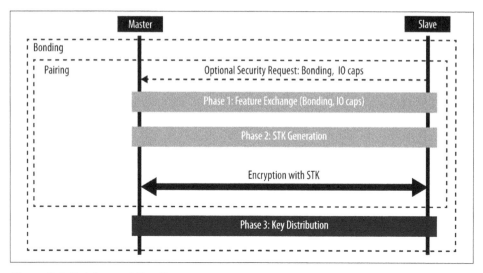

Figure 2-6. Pairing and Bonding sequences

Pairing Algorithms

A pairing procedure involves an exchange of Security Manager Protocol (SMP) packets to generate a temporary encryption key called Short Term Key (STK) on both sides. The last step of a *pairing* procedure (regardless of whether it will then continue into a security key exchange and therefore a *bonding* procedure) is to encrypt the link with the previously generated STK. During the packet exchange, the two peers negotiate one of the following STK generation methods:

Just Works
> The STK is generated on both sides, based on the packets exchanged in plain text. This provides no security against *man-in-the-middle* (MITM) attacks.

Passkey Display
> One of the peers displays a randomly generated, six-digit passkey and the other side is asked to enter it (or in certain cases both sides enter the key, if no display is available). This provides protection against MITM attacks and is used whenever possible.

Out Of Band (OOB)
> When using this method, additional data is transferred by means other than the BLE radio, such as another wireless technology like NFC. This method also provides protection against MITM attacks.

The SM specifies the following three types of security mechanisms that can be used to enforce various levels of security while in a connection or during the advertising procedure:

Encryption
> This mechanism consists of the full encryption of all packets transmitted over an established connection.

Privacy
> The privacy feature allows an advertiser to hide its public Bluetooth address by using temporary, randomly generated addresses that can be recognized by a scanner that is bonded with the advertising device.

Signing
> With this mechanism, a device can send an unencrypted packet over an established connection that is digitally signed (i.e., the source of which can be verfied).

Each of these three mechanisms can be used independently from the others, and the application, in conjunction with the host, has the choice of enforcing any of them concurrently.

Security Keys

Each of the three types of security mechanisms requires a key or a group of keys to be established. These are the keys that are exchanged and stored to allow for the security mechanisms to be enforced:

Encryption Information (Long Term Key or LTK) and Master Identification (EDIV, Rand)
> This is a 128-bit encryption key shared by both sides (LTK) along with two values (EDIV, Rand) acting as its identifier, since a device may be bonded with multiple other peers.

Identity Information (Identity Resolving Key or IRK) and Identity Address Information (Address Type and Bluetooth Device Address)
> The IRK is used to implement privacy: it can generate and resolve resolvable addresses (see "Address Types" on page 44) that protect an advertising device from being tracked by malicious peers. The actual public or static random address of the device distributing it is included along with the IRK.

Signing Information (Connection Signature Resolving Key or CSRK)
> A key used to digitally sign unencrypted data.

Each of these keys or key sets is *asymmetrical* and *unidirectional*: it can be used only in the same role configuration in which they were originally generated. If the devices wish to switch Link Layer roles (master and slave) in subsequent connections, then each side must distribute its own set of keys for each key type. Devices negotiate the number of keys distributed in each direction, which can range from zero to all three key types in each direction, for a total of six keys distributed between peers (three from slave to master and three from master to slave).

If no keys at all are exchanged during a bonding procedure, the bond between the two devices will still be valid, but no security procedures whatsoever will be available between them.

Since each key is asymmetrical (and therefore the process of key distribution is symmetrical) and thus each bond information stored between two devices can contain up to two instances of each key (each peer having distributed its own), it's important to note how keys distributed by each device are used in subsequent connections. Table 2-2 details this usage based on the distributors and acceptors during the bonding procedure.

Table 2-2. Security Manager key usage

Key	Distributor Usage	Acceptor Usage
LTK, EDIV, Rand (Encryption)	Used to encrypt the link when a slave	Used to encrypt the link when a master
IRK, BD_ADDR (Privacy)	Used to generate resolvable private addresses	Used to resolve resolvable private addresses
CSRK (Signing)	Used to sign data	Used to verify signatures

As an example of a bonding with key distribution, let's assume that two devices, a tablet acting as a master and a watch acting as a slave, perform a bonding procedure and exchange encryption keys in both directions. The watch will distribute its own encryption keys in the form of encryption information and master identification (let's call them `LTK_EDIV_Rand_watch`) and the tablet will do the same thing in the opposite direction (`LTK_EDIV_Rand_tablet`).

After the bonding is complete, the link can be disconnected, and then the two devices might want to reconnect and reuse those keys to reestablish a secure, encrypted connection without having to go through the bonding procedure again. If the devices reconnect in the same configuration as before, with the tablet acting as a master, then both will use `LTK_EDIV_Rand_watch` to encrypt the link. If, later, the two reconnect with switched roles (i.e., the watch is this time acting as the master and the table a slave), `LTK_EDIV_Rand_tablet` can then be used to encrypt the link.

Chapter 3 expands on these concepts in more detail.

Generic Attribute Profile (GATT)

The Generic Attribute Profile (GATT) builds on the Attribute Protocol (ATT) and adds a hierarchy and data abstraction model on top of it. In a way, it can be considered the backbone of BLE data transfer because it defines how data is organized and exchanged between applications.

It defines generic data objects that can be used and reused by a variety of application profiles (known as *GATT-based profiles*). It maintains the same client/server architecture present in ATT, but the data is now encapsulated in *services*, which consist of one or more *characteristics*. Each characteristic can be thought of as the union of a piece of user data along with *metadata* (descriptive information about that value such as properties, user-visible name, units, and more).

Chapter 4 discusses GATT in much more detail. Along with GAP, GATT is an upper-layer entity that acts as the main interface to a Bluetooth Low Energy protocol stack.

Generic Access Profile (GAP)

The Generic Access Profile (GAP) dictates how devices interact with each other at a lower level, outside of the actual protocol stack. GAP can be considered to define the BLE topmost control layer, given that it specifies how devices perform control procedures such as device discovery, connection, security establishment, and others to ensure interoperability and to allow data exchange to take place between devices from different vendors.

GAP establishes different sets of rules and concepts to regulate and standardize the low-level operation of devices:

- Roles and interaction between them
- Operational modes and transitions across those
- Operational procedures to achieve consistent and interoperable communication
- Security aspects, including security modes and procedures
- Additional data formats for nonprotocol data

Chapter 3 discusses GAP in more detail.

GAP (Advertising and Connections)

The Generic Access Profile (GAP) is the cornerstone that allows Bluetooth Low Energy devices to interoperate with each other. It provides a framework that any BLE implementation must follow to allow devices to discover each other, broadcast data, establish secure connections, and perform many other fundamental operations in a standard, universally understood manner. It's important to understand GAP thoroughly, because many BLE protocol stacks use it as one of the lowest-level entry points when providing a functional API to application developers.

As mentioned previously, the sections of the GAP chapter in the core specification that apply to Bluetooth Low Energy define the following different aspects of device interaction:

Roles

Each device can operate in one or more roles at the same time. Each role imposes restrictions and enforces certain behavioral requirements. Certain combinations of roles allow devices to communicate with each other, and GAP establishes the interactions between those roles precisely. Although not always, roles tend to be associated with specific device types, and for many (though not all) implementations, they are also tightly bound with their use case and do not change at all.

Modes

Further refining the concept of a role, a mode is a state in which the device can switch to for a certain amount of time to achieve a particular goal or, more especifically, to allow a peer to perform a particular procedure. Switching modes can be triggered by user interface actions or automatically when required, and devices tend to switch modes more frequently than roles.

Procedures

A procedure is a sequence of actions (usually Link Layer control sequences or packet exchanges) that allows a device to attain a certain goal. A procedure is typically associated with a mode on the other peer, so they are often tightly coupled together.

Security

GAP builds on top of the Security Manager and the Security Manager Protocol by defining security modes and procedures that specify how peers set the level of security required by a particular data exchange and later how that security level is enforced. GAP further defines additional security features not associated with particular modes or procedures, and implementations are free to use those to increase the level of data protection required by each application.

Additional GAP Data Formats

In addition to all of the above, GAP is also used as a placeholder for certain additional data format definitions that are related to the modes and procedures defined by the GAP sepcification.

The corresponding sections in this chapter will examine these elements in detail.

Roles

GAP specifies four roles that a device can adopt to join a BLE network:

Broadcaster

Optimized for transmit-only applications that distribute data regularly, the *broadcaster* role periodically sends out advertising packets with data. Theoretically, the broadcaster role could be used with transmitter-only radios, but in practice, this role is usually assigned to a device capable of both transmitting and receiving. A public thermometer that broadcasts temperature readings to any interested devices would be a good example of a broadcaster. Broadcasters send data in advertising packets rather than connection data packets, and the data is accessible to any device that is listening. The broadcaster role uses the Link Layer advertiser role.

Observer

Optimized for receive-only applications that want to collect data from broadcasting devices, the *observer* role listens for data embedded in advertising packets from broadcasting peers. For example, a device with a display is a typical application of this role, such as a table computer that displays temperature data from a broadcast-only temperature sensor. The observer role uses the Link Layer scanner role.

Central

The central role corresponds to the Link Layer master. A device capable of establishing multiple connections to peers, the *central* role is always the initiator of connections and essentially allows devices onto the network. The BLE protocol is *asymmetric*, which means that the computing requirements of the Link Layer master are larger than the ones of a Link Layer slave. The central role is usually played by a smartphone or tablet in the network, because it has access to powerful CPUs and memory resources. This allows it to maintain connections to multiple devices. The central starts by listening for other devices' advertising packets and then initiates a connection with a selected device. This process can be repeated to include multiple devices in a single network.

Peripheral

The peripheral role corresponds to the Link Layer slave. This role uses advertising packets to allow centrals to find it and, subsequently, to establish a connection with it. The BLE protocol is optimized to require few resources for peripheral implementation, at least in terms of processing power and memory. This paves the way to a large market of inexpensive BLE peripherals.

Each particular device can operate in one or more roles at a time, and the specification imposes no restrictions on this regard.

Many developers mistakenly try to associate the BLE GATT *client* and *server* roles with GAP roles. There is no connection between those at all, and any device can be a GATT client, server, or both, depending on the application and situation.

Consider, for example, a fitness tracker paired with a smartphone. The fitness tracker's GAP role is *peripheral*, and it acts as a GATT server when the phone requests data from its sensors. It can also sometimes act as a GATT client when it requests accurate time data from the smartphone to update its internal clock for data timestamping. The GATT client/server roles depend exclusively on the direction in which the data requests and responses transactions flow, whereas GAP roles stay constant as *peripheral* for the fitness tracker and *central* for the smartphone.

Modes and Procedures

Table 3-1 shows GAP modes and their applicable procedures (modes that do not have a natural counterpart procedure are marked with "N/A").

Table 3-1. Modes and their applicable procedures

Mode	Applicable Role(s)	Applicable Peer Procedure(s)
Broadcast	Broadcaster	Observation
Non-discoverable	Peripheral	N/A
Limited discoverable	Peripheral	Limited and General discovery
General discoverable	Peripheral	General discovery
Non-connectable	Peripheral, broadcaster, observer	N/A
Any connectable	Peripheral	Any connection establishment

Conversely, Table 3-2 shows the modes that the peer needs be in to perform each of the listed GAP procedures.

Table 3-2. Procedures and their required modes

Procedure	Applicable Role(s)	Applicable Peer Mode(s)
Observation	Observer	Broadcast
Limited discovery	Central	Limited discoverable
General discovery	Central	Limited and General discoverable
Name discovery	Peripheral, central	N/A
Any connection establishment	Central	Any connectable
Connection parameter update	Peripheral, central	N/A
Terminate connection	Peripheral, central	N/A

Chapter 1 and Chapter 2 introduced the basic concepts of over-the-air data exchange in BLE, but it is worth reviewing them briefly here. Advertising packets are blindly sent unidirectionally at fixed intervals, and they constitute the basis of both broadcasting (and observing) and discovery. A device scanning for advertising packets might receive one if it happens to scan while an advertising packet is being transmitted, and it might simply receive the data contained in it or continue by initiating a connection. *Connections*, on the other hand, require two peers that synchronously perform data exchanges at regular intervals and provide guarantees on data transmission and throughput.

Broadcast and Observation

The *broadcast* mode and the *observation* procedure defined in GAP establish the framework through which devices can send data unidirectionally, as a *broadcaster* to one or more actively listening peer devices (the *observers*). It is important to note that the broadcaster has no way of knowing whether the data actually reaches any observers at all, so this combination of mode and procedure remains faithful to its nomenclature: a broadcaster broadcasts data without any confirmation or acknowledgement, and an observer listens (temporarily or indefinitely) for potential broadcasters without any guarantee of ever actually receiving any data.

The advertising packets sent by the broadcaster contain actual valid user data, along with a few items of metadata (such as Bluetooth device address) inserted by the Link Layer. As described in "Advertising and Scanning" on page 19, each advertising packet contains up to 31 bytes of data (the actual available user data length will be lower due to headers and format overheads), but that can be doubled by using the scan request/ scan response transaction just after the successful reception of an advertising packet on the part of the observer, yielding up to 62 bytes of data per advertising event. Since a scan response packet is sent only upon request from the observer, the most critical and important data should always be placed in the advertising packet itself, not in the scan response packet. A broadcaster can send `ADV_NONCONN_IND` or `ADV_SCAN_IND` advertising packets (see Table 2-1).

By creating a broadcaster-only device, you can simply *broadcast* data to the outside world, where any device within listening range can pick it up, whether that means one device or one hundred devices. This is in marked contrast to a *peripheral*, which stops advertising itself after establishing a connection, effectively shutting itself off to any other central devices in listening range until the connection is closed or, in the rare case of devices that support multiple connections as a slave, until an additional connection is created.

For example, Apple's iBeacon ("iBeacon" on page 132) uses the broadcast mode to constantly send out a specific payload in the Manufacturer Specific Data field of the advertising data, which allows any device that comes within earshot of the node to detect the iBeacon without having to compete for access with other devices in range. The iBeacon node doesn't have to worry about how many devices are listening; it just keeps telling the world that it's there and transmitting its limited payload to anyone who cares to listen.

Discovery

A device's *discoverability* refers to how the peripheral advertises its presence to other devices and what those devices can or should do with that information. The differences between the different discoverable modes and discovery procedures concern whether advertising and scanning are actually being performed but also take into account the nature of the data included in advertising packets. More specifically, a SIG-defined optional field within the advertising data named *Flags AD* governs a device's discoverable mode (see Table 3-3).

Used only by peripherals, these modes allow central devices to discover peripherals within their listening range. *Discovery* commonly refers to detecting the presence and the basic information of another device nearby. That does not necessarily imply an intention to create a connection or exchange data, although that is naturally often the case. In some instances, and especially with central devices equipped with user-visible displays, discovery is simply used to populate a list with nearby devices from which the user can then select one.

Discoverability modes

The following discoverability modes allow a certain amount of flexibility to peripheral designers, depending on design priorities (battery life, fast connection times, etc.):

Non-discoverable mode
    Not being discoverable means other devices cannot learn about the presence of the peripheral or perform any inquiries about its nature. This mode is typically used

when a device does not want to be found by central peers at all, either to establish a connection or even to be detected (listed). A device in this mode can still send advertising packets, but in that case, the device must clear the General and Limited Discoverable bits in the Flags AD field within its advertising data. If it does choose to send advertising packets, they must be of the ADV_NONCONN_IND or ADV_SCAN_IND types (see Table 2-1).

Limited discoverable mode

This mode allows a device to be discoverable for a limited period of time and with a lowered priority. A device in this mode sends advertising packets with the Limited Discoverable flag in the Flags AD field set (see Table 3-3). Central devices performing the limited discovery procedure will be able to detect only devices in this mode. The popularity of this mode has been dwindling with time, and today the tendency is to use the general discoverable mode with additional filters when required.

General discoverable mode

This mode makes a device discoverable for as long as it is required or deemed necessary. A device switching to this mode expresses its desire to be discovered by central peers, generally with the intention of establishing a connection. To be in this mode, a device must set the General Discoverable flag in the Flags AD field of the advertising data while sending advertising packets (see Table 3-3). Only central devices performing the general discovery procedure will find peripherals in this mode.

Peripheral devices often come up from the factory in a discoverable mode before bonding with a central device, but then go into non-discoverable mode after the initial bonding procedure, allowing it to connect exclusively with that central device in the future. In this case, a reset to factory defaults usually brings it back into a discoverable mode.

Discovery procedures

The specification provides two discovery procedures:

Limited discovery procedure

A central performing this procedure starts active scanning with no white list filtering (see "White Lists" on page 23) for advertisers and analyzes each advertising packet that it receives. If the Limited Discoverable flag is set, the peer device is reported to the application for further action.

General discovery procedure

A central performing this procedure starts active scanning with no white list filtering (see "White Lists" on page 23) for advertisers and analyzes each advertising packet that it receives. If either the Limited Discoverable or General Discoverable flags are set, the peer device is reported to the application for further action.

In practical terms, devices looking for all possible discoverable peers should opt for the general discovery procedure. Use the limited discovery procedure to find only devices in limited discovery mode.

Connection Establishment

For a central to initiate a connection establishment with a peripheral, the latter must be in a *connectable* mode. Similar to discovery, several modes and procedures control the selection of devices with which to interact, in an organized and standardized fashion.

Connection establishment modes

The differences between the following connection establishment modes reflect a peripheral's use of different types of advertising packets (detailed in "Advertising and Scanning" on page 19):

Non-connectable mode
> A device in this mode either does not send advertising packets at all or it sends `ADV_NONCONN_IND` or `ADV_SCAN_IND` advertising packets (see Table 2-1). In both cases, the device is, as the mode name implies, not connectable, meaning that no centrals may establish a connection with it.

Directed connectable mode
> A device in this mode sends `ADV_DIRECT_IND` advertising packets (see Table 2-1). When performing directed advertising, a device sends advertising packets at a high frequency and for a short time, with no user data payload and with a target central Bluetooth Address. This is provided as a "fast reconnect" mode typically used when the peripheral has a strong suspicion that the target central is already trying to initiate a connection and wants to establish it as fast as possible. Only the central whose Bluetooth address matches the one in the advertising packets sent by the peripheral will receive them.

Undirected connectable mode
> A device in this mode sends `ADV_IND` advertising packets (see Table 2-1). This is the standard connectable mode, through which a peripheral makes itself connectable for a longer period of time and may be trying to connect to a new central or to one that is already previously known by it.

Both connectable modes implicity require the device to send the advertising packets with the intent to connect to a central.

Connection establishment procedures

Because a central device has no means to select which advertising packet types it will receive when scanning with the intent to connect (they will always be of type `ADV_IND` or `ADV_DIRECT_IND`), the distinctions between connection establishment procedures do not depend on types of advertising packets. Instead, the type of connection establishment procedure used depends on the kind of filtering the central imposes on those incoming packets:

Auto connection establishment procedure
> With this single-step procedure, the host populates a white list (see "White Lists" on page 23) with an array of known peripheral devices and then instructs the controller to connect to the first one detected. Generally speaking, this procedure is useful when the central already knows a limited set of devices and does not have a preference for which one to connect to.

General connection establishment procedure
> This two-step procedure is commonly used to connect to a new, unknown peripheral. The central starts by scanning without a white list, accepting all incoming advertising packets. For each peripheral detected, the application then needs to decide whether to connect to it or to continue to the next one. To do so, the application can prompt the user or parse advertising data included in the received advertising packet. Once a peripheral is chosen, the central connects to it using the direct connection establishment procedure.

Selective connection establishment procedure
> This procedure is identical to the general connection establishment procedure, with the exception that the host uses a white list of previously known devices to filter incoming advertising packets. This can be useful in certain cases in which the user should choose which of several known peripherals to connect to.

Direct connection establishment procedure
> This standard single-step connection establishment procedure connects the central to one particular peripheral. The host uses the Link Layer to initiate a connection to one single device, identified by its Bluetooth Address, without previous knowledge of its presence. The procedure can fail if the targeted peripheral is not available or is not in a connectable mode.

It is worth reiterating that a central host has two different ways to initiate a connection. The first method requires two steps: first scanning and then connecting directly to a device (by specifying its Bluetooth Address) detected during the scanning phase. The second method skips the explicit scanning step and instead uses the controller to select one or more devices to connect to, without previous knowledge of whether they are actually nearby.

Additional GAP Procedures

GAP defines a few other procedures that are relevant only to already established connections, and which are commonly used:

Name discovery procedure

Advertising packets can carry many different types of user data, including the *Device Name*: a UTF-8 string containing a human-readable description of the device (similar to a network hostname). But space in the advertising packet is at a premium, so sometimes devices might not choose to include the Device Name. For such cases, the name discovery procedure allows a peripheral or central to retrieve the Device Name over an established connection by using a GATT transaction.

Connection parameter update procedure

Each connection establishment requires a set of connection parameters that are set unilaterally by the central. These parameters are the key factors in the balance between throughput and power consumption for a connection (as described in "Connections" on page 22) and can be modified later in a connection to adapt to changing balance requirements. The central is always responsible for physically modifying the connection parameters and can do so at any time without warning, but the peripheral also has the ability to *request* the central for a change to a specific desired set of connection parameters. After doing so, the central can opt to deny the request or to honor it. However, even if it does decide to change the connection parameters, they might not be the exact ones requested by the peripheral, but rather whatever the central considers reasonable and closest to the requested set.

Terminate connection procedure

This procedure is self explanatory and completely symmetrical: both the peripheral and the central can terminate the connection at any time, while providing a *reason* code that the peer application will receive along with the disconnection event.

Although this concludes the general GAP modes and procedures portion of this chapter, "Security" introduces specific modes and procedures exclusive to the subject of security.

Security

Security is built into the core of the Bluetooth Low Energy wireless technology. Since the introduction of BLE in the Bluetooth 4.0 core specification, the secure transmission of user data has been a primary design goal, and version 4.1 builds upon the strong foundations of its predecessor and tightens those principles even further.

As introduced in "Security Manager (SM)" on page 28, the enforcement of security and privacy in BLE is based on two pillars:

Security Manager (and its protocol)

The Security Manager implements the actual cryptographic algorithms and protocol exchanges that allow two devices to securely exchange data and privately detect each other. This is typically implemented inside the protocol stack and uses the available hardware acceleration modules to perform the required operations, such as generating random numbers, encrypting data using the Advanced Encryption Standard (AES), and generating and exchanging security keys with the peer.

GAP security aspects

GAP defines a set of modes and procedures related to security that allow the implementation to trust the connection to carry sensitive data and to confidently accept the identity of a peer device. Security procedures are mostly asymmetrical, with the central and the peripheral assuming different roles during the generation and exchange of keys and the subsequent use of them to establish a secure connection. Additionally, GAP further expands and specifies the usage of the tools defined in "Security Manager (SM)" on page 28 in a standard and interoperable manner.

The following sections further detail the GAP security aspects that, although at times complex and even convoluted, are fundamental to the operation of a majority of BLE devices.

Address Types

As discussed in "Bluetooth Device Address" on page 19, at its lowest level, the BLE protocol stack differentiates between public and random device addresses. GAP extends the concept of random device addresses by classifying them into three different categories:

Static address

Static addresses are typically used as a replacement for public ones whenever the manufacturer does not want or need the overhead of IEEE registration. A static address is simply a random number that can either be generated every time the device boots up or can stay the same for the lifetime of the device. However, it cannot be changed during a power cycle of the device.

Non-resolvable private address

This type of address is not commonly used. Also a randomly generated number, it represents a temporary address used for a certain amount of time.

Resolvable private address

Resolvable private addresses form the basis of the privacy feature. They are generated from an identity resolving key (IRK) and a random number, and they can be changed often (even during the lifetime of a connection) to avoid the device being identified and tracked by an unknown scanning device. Only devices that possess the IRK distributed by the device using a private resolvable address can actually resolve that address, allowing them to identify the device.

GAP also defines an encoding scheme to obtain the category from the two highest bits of the BD_ADDR 48-bit value, which can be useful in certain situations. But to ascertain the nature of a BLE address, you need one extra bit on top of the 48-bit, because otherwise it would be impossible to know whether the BD_ADDR corresponds to a public or a random device address (public device addresses make no guarantees regarding the contents of its most significant bits).

Authentication

The usage of the word *authentication* and its derivatives in the BLE specification (and in GAP in particular) is rather convoluted and can lead to confusion, so this section attempts to clarify the two different meanings it might convey:

Authentication as a procedure
In GAP, the authentication procedure refers to *enforcing security on the current connection*. This can refer to pairing (with or without bonding) or re-establishing security using a stored encryption key, as described in "Security Procedures" on page 29. The verb *to authenticate* is used in the same way (i.e., to perform the authentication procedure).

Authenticated as a qualifier
As a qualifier for a pairing procedure or an existing key, *authenticated* means *using an algorithm other than Just Works* (see "Pairing Algorithms" on page 30). In this sense, it implies that MITM protection was available during the pairing procedure and key exchange, adding an additional level of trust. Both passkey display and out-of-band (OOB) are algorithms that produce authenticated keys. Conversely, an unauthenticated pairing procedure (using Just Works) can produce only unauthenticated keys. When applied to encryption or data signing, *authenticated* refers to the keys used to perform those procedures.

GAP APIs and documentation use both meanings extensively, so it is important to keep the differences in mind for the next few sections.

Security Modes

A connection is said to operate in a (single) *security mode*. The use of *mode* in this context differs sharply from the one in previous sections. This defines the current security level of the connection, which at some point in time might differ from the requirements that the peers at each end want to enforce, leading to procedures to increase that security level.

GAP defines two security modes, along with several security levels per mode:

Security mode 1
This mode enforces security by means of encryption, and it contains three levels:

Level 1
No security (the link is not encrypted).

Level 2
Unauthenticated encryption.

Level 3

Authenticated encryption.

Security mode 2

This mode enforces security by means of data signing (see "Security Manager (SM)" on page 28), and it contains two levels:

Level 1

Unauthenticated data signing.

Level 2

Authenticated data signing.

Each connection starts its lifetime in security mode 1, level 1, and can later be upgraded to any of the security modes by means of encryption or data signing. It is important to know that a link can be downgraded from mode 1, level 3, to mode 1, level 2 by switching encryption keys, but encryption can never be disabled in the lower layers, making it impossible to go down from security mode 1, level 2.

Security Modes and Procedures

Along with all the modes and procedures detailed in previous sections, GAP also defines additional ones related to security establishment and enforcement.

 In this section, the term *mode* refers back to the temporary state to which a device can switch in order to perform a procedure or to allow a procedure to be performed.

This section briefly describes the security modes and procedures, which complement and build upon the basis set in "Security Manager (SM)" on page 28.

Non-bondable mode

A device in this mode does not allow a bonding procedure to take place, although it can freely permit pairing procedures to execute. In this mode, a device cannot distribute, accept, or store keys, limiting all security level upgrades to the lifetime of the connection.

Bondable mode

This mode enables the device to create a bond with a peer, permanently storing security keys.

Bonding procedure

The central can initiate the bonding procedure described in "Security Procedures" on page 29 at any time (even if the devices are already bonded, in which cases new keys would be generated and the old ones would be replaced). However, it is relatively common practice to hold off until a GATT data access actually requires a higher security mode than the one currently in force.

Authentication procedure

As mentioned in "Authentication" on page 45, the authentication procedure enforces the transition to a security mode on a connection currently in a lower security mode. This enforcement can come in the form of a pairing or bonding procedure or, if encryption keys are already available and sufficiently authenticated on both sides, in the form of an encryption procedure.

Authorization procedure

Authorization in BLE and GAP refers to giving the application and, ultimately, the user the opportunity to accept or reject a particular transaction. This might take the shape of an internal application check if the device does not have a user interface, or if it does, it can ask the user for permission directly.

Encryption procedure

This procedure uses the Link Layer's built-in encryption capabilities to start the encryption of all traffic over the current connection, as described in "Connections" on page 22 and "Security Procedures" on page 29. Although the central is the only one able to initiate an encryption procedure, the peripheral can request the central to do so by means of the *security request* mentioned in "Security Manager (SM)" on page 28.

Although not strictly a mode or procedure, the *privacy* feature deserves a few words in the section on GAP security. Privacy was one of the aspects of the BLE specification that underwent a substantial modification between versions 4.0 and 4.1. The feature was simplified and the specification text was adapted to reflect what implementations had been using in practice for years, acommodating itself to the established status quo among currently shipping devices.

This revision was named *Privacy 1.1* and was actually listed as a new feature in the 4.1 release, replacing the old privacy section. When using the privacy feature, a device periodically generates a new resolvable private address (see "Address Types" on page 44) and sets it as its own. The only way for a peer to identify it is to resolve the address using an IRK, because all resolvable private addresses are temporary in nature and it makes no sense to store them permanently.

Instead, an IRK (see "Security Keys" on page 31) previously shared and stored by both devices is used to generate the address in the device, protecting its identity and resolving it on the device trying to identify it. This feature can be used in all GAP roles concurrently and prevents malicious devices deprived of the shared IRK from identifying and tracking a particular device using its Bluetooth device address.

Additional GAP Definitions

On top of all the roles, modes, and procedures, GAP also introduces two additional items that are relevant to the interests of application developers.

Advertising Data Format

So far, we have been talking about the user data that can be carried by advertising (and scan response) packets, but we have not mentioned the format in which this data must be populated. The generic container is defined in the core specification in GAP, and it simply consists of a sequence of data structures, each of which is made of length (1 byte), *AD Type* (Advertising Data Type, 1 byte), and the actual data (variable length). Each structure contains a separate item of user data.

The complete list of allowed AD Types is available in the Core Specification Supplement (CSS, currently at version 4) at the Bluetooth SIG's Specification Adopted Documents page (*http://bit.ly/1gg4UhU*). Table 3-3 describes the most commonly used AD Types in typical application development. This list isn't exhaustive, though, so you might want to explore AD Types further for more detailed information.

Table 3-3. AD Types

Name	Actual data length in bytes	Description
Flags	1 (extendable)	Used to set limited or general discovery mode, as described in "Discovery" on page 39
Local Name	variable	Partial or complete user-readable local name in UTF-8
Appearance	2	A 16-bit value describing the type of device sending the advertising packet
TX Power Level	1	The power level in dBm used to transmit the advertising packet, useful to calculate path loss at the observer or central end
Service UUID	variable	A complete or partial list of GATT services offered by the device sending the packet (as a GATT server)
Slave Connection Interval Range	4	A suggestion to the central about the connection interval range that best fits this peripheral
Service Solicitation	variable	A list of GATT services supported by the device sending the packet (as a GATT client)
Service Data	variable	A UUID representing a GATT service and its associated data
Manufacturer Specific Data	variable	Freely formattable data, to be used at the discretion of the implementation

Not all AD Types are useful in every situation, and the 31-byte advertising packet and scan reponse size limits how many AD Types you can include, but it's important to understand what types of information are available and which fields are relevant in different situations. This section briefly describes how the service-related AD Types can be useful for a wide range of applications.

Services and service UUIDs are used to encapsulate data into logic groups in BLE (see Chapter 4 for more details). The Bluetooth SIG defines a number of official services, all of which have a unique UUID associated with them, and you are free to create your own custom services with their own dedicated UUID. Services are what make device interoperability possible. For example, any device that implements the Battery Service must do so based on the same data contract defined by the Bluetooth SIG, using the same UUIDs and data format.

An advertising packet can include complete or incomplete lists of service UUIDs under the Service UUID AD Type to let other devices know which services are exposed by the peripheral without having to establish a connection with it. Connections are expensive, and limiting the occasions where connections are necessary can help extend battery life. By letting the world know that the device exposes a certain set of services as a GATT server, you can more quickly determine if that device is relevant to you without connecting to every device that comes within range.

Conversely, service solicitation allows the central receiving the advertising packets to find out which services the peripheral can access when it is acting as a GATT client. In certain cases, if the central is not exposing any of these services at all as a GATT server, it might not even bother to connect to it, knowing that interaction would be extremely limited.

While services are normally used only after a connection is established, the advertising packet includes a mechanism to offer service data right in the advertising payload: the Service Data. The key benefit of using the Service Data field is that it allows you to broadcast the data contained inside your GATT services to any number of listening devices, bypassing the need to a dedicated connection between two devices to read service data.

Finally, the Manufacturer Specific Data AD Type is a generic catch-all field that can include an arbitrary chunk of data in your advertising payload. For example, Apple's iBeacon ("iBeacon" on page 132) uses this AD type to push data out to other devices, inserting a 128-bit UUID that identifies the location and two 16-bit values to distinguish individual nodes in the same family. This field provides a lot of flexibility to product designers who want to broadcast short chunks of custom data, and you're generally free to change to contents of the field from one advertising event to the next.

Please note that the Bluetooth SIG might release further updates to the Core Specification Supplement at the Specification Adopted Documents page (*http://bit.ly/ 1gg4UhU*) whenever new AD Types are available for public consumption.

GAP Service

The final item that GAP includes as part of its section in the core specification is the GAP Service, a mandatory GATT service that every device must include among its attributes (see "Attribute Protocol (ATT)" on page 26). The service is freely accessible (read-only) to all connected devices with no security requirements whatsoever, and it contains the following three characteristics:

Device Name characteristic

This contains the same user-readable UTF-8 string that can also be included in the Device Name AD Type described in Table 3-3. This is the characteristic read to perform the name discovery procedure in "Additional GAP Procedures" on page 42.

Appearance characteristic

This 16-bit value associates the device in which it is included with a certain generic category (phone, computer, watch, sensor, etc.) and is typically used by the GATT client to display an icon that represents the given category. This characteristic can also be made available in the advertising packet with the Appearance AD Type.

Peripheral Preferred Connection Parameters (PPCP) characteristic

Once the central has established a connection with the peripheral, it can (although it has no obligation to) then read the value of this characteristic and perform a connection parameter update procedure (described in "Additional GAP Procedures" on page 42) to change the connection's parameters to values that are agreeable to the peripheral.

Chapter 4 explores GATT services and characteristics in more detail, along with all of the procedures that allow BLE devices to exchange user data over an established connection.

GATT (Services and Characteristics)

The Generic Attribute Profile (GATT) establishes in detail how to exchange all profile and user data over a BLE connection. In contrast with GAP (Chapter 3), which defines the low-level interactions with devices, GATT deals only with actual data transfer procedures and formats.

GATT also provides the reference framework for all *GATT-based profiles* (discussed in "SIG-defined GATT-based profiles" on page 14), which cover precise use cases and ensure interoperability between devices from different vendors. All standard BLE profiles are therefore based on GATT and must comply with it to operate correctly. This makes GATT a key section of the BLE specification, because every single item of data relevant to applications and users must be formatted, packed, and sent according to its rules.

GATT uses the Attribute Protocol (detailed in "Attribute Protocol (ATT)" on page 26) as its transport protocol to exchange data between devices. This data is organized hierarchically in sections called *services*, which group conceptually related pieces of user data called *characteristics*. This determines many fundamental aspects of GATT discussed in this chapter.

Roles

As with any other protocol or profile in the Bluetooth specification, GATT starts by defining the roles that interacting devices can adopt:

Client

The GATT client corresponds to the ATT client discussed in "Attribute Protocol (ATT)" on page 26. It sends requests to a server and receives responses (and server-initiated updates) from it. The GATT client does not know anything in advance about the server's attributes, so it must first inquire about the presence and nature of those attributes by performing service discovery. After completing service discovery, it can then start reading and writing attributes found in the server, as well as receiving server-initiated updates.

Server

The GATT server corresponds to the ATT server discussed in "Attribute Protocol (ATT)" on page 26. It receives requests from a client and sends responses back. It also sends server-initiated updates when configured to do so, and it is the role responsible for storing and making the user data available to the client, organized in attributes. Every BLE device sold must include at least a basic GATT server that can respond to client requests, even if only to return an error response.

It is worth mentioning once more that GATT roles are both completely independent of GAP roles (see "Roles" on page 36) and also concurrently compatible with each other. That means that both a GAP central and a GAP peripheral can act as a GATT client or server, or even act as both at the same time.

UUIDs

A universally unique identifier (UUID) is a 128-bit (16 bytes) number that is guaranteed (or has a high probability) to be globally unique. UUIDs are used in many protocols and applications other than Bluetooth, and their format, usage, and generation is specified in ITU-T Rec. X.667 (*http://bit.ly/1gScfna*), alternatively known as ISO/IEC 9834-8:2005 (*http://bit.ly/1es3R3t*).

For efficiency, and because 16 bytes would take a large chunk of the 27-byte data payload length of the Link Layer, the BLE specification adds two additional UUID formats: 16-bit and 32-bit UUIDs. These shortened formats can be used only with UUIDs that are defined in the Bluetooth specification (i.e., that are listed by the Bluetooth SIG as standard Bluetooth UUIDs).

To reconstruct the full 128-bit UUID from the shortened version, insert the 16- or 32-bit short value (indicated by xxxxxxxx, including leading zeros) into the Bluetooth Base UUID:

```
xxxxxxxx-0000-1000-8000-00805F9B34FB
```

The SIG provides (shortened) UUIDs for all the types, services, and profiles that it defines and specifies. But if your application needs its own, either because the ones offered by the SIG do not cover your requirements or because you want to implement a new use case not previously considered in the profile specifications, you can generate them using the ITU's UUID generation page (*http://bit.ly/1kQOvDM*).

Shortening is not available for UUIDs that are not derived from the Bluetooth Base UUID (commonly called *vendor-specific UUIDs*). In these cases, you'll need to use the full 128-bit UUID value at all times.

Attributes

Attributes are the smallest data entity defined by GATT (and ATT). They are *addressable* pieces of information that can contain relevant user data (or *metadata*) about the structure and grouping of the different attributes contained within the server. Both GATT and ATT can work only with attributes, so for clients and servers to interact, all information must be organized in this form.

Conceptually, attributes are always located on the server and accessed (and potentially modified) by the client. The specification defines attributes only conceptually, and it does not force the ATT and GATT implementations to use a particular internal storage format or mechanism. Because attributes contain both static definitions of invariable nature and also actual user (often sensor) data that is bound to change rapidly with time (as discussed in "Attribute and Data Hierarchy" on page 56), attributes are usually stored in a mixture of nonvolatile memory and RAM.

Each and every attribute contains information about the attribute itself and then the actual data, in the fields described in the following sections.

Handle

The attribute handle is a unique 16-bit identifier for each attribute on a particular GATT server. It is the part of each attribute that makes it addressable, and it is guaranteed not to change (with the caveats described in "Attribute Caching" on page 66) between transactions or, for bonded devices, even across connections. Because value 0x0000 denotes an invalid handle, the amount of handles available to every GATT server is 0xFFFF (65535), although in practice, the number of attributes in a server is typically closer to a few dozen.

> Whenever used in the context of attribute handles, the term *handle range* refers to all attributes with handles contained between two given boundaries. For example, handle range 0x0100-0x010A would refer to any attribute with a handle between 0x0100 and 0x010A.

Within a GATT server, the growing values of handles determine the ordered sequence of attributes that a client can access. But gaps between handles are allowed, so a client cannot rely on a contiguous sequence to guess the location of the next attribute. Instead, the client must use the discovery feature ("Service and Characteristic Discovery" on page 69) to obtain the handles of the attributes it is interested in.

Type

The attribute type is nothing other than a UUID (see "UUIDs" on page 52). This can be a 16-, 32-, or 128-bit UUID, taking up 2, 4, or 16 bytes, respectively. The type determines the kind of data present in the value of the attribute, and mechanisms are available to discover attributes based exclusively on their type (see "Service and Characteristic Discovery" on page 69).

Although the attribute type is always a UUID, many kinds of UUIDs can be used to fill in the type. They can be standard UUIDs that determine the layout of the GATT server's attribute hierarchy (further discussed in "Attribute and Data Hierarchy" on page 56), such as the *service* or *characteristic* UUIDs, profile UUIDs that specify the kind of data contained in the attribute, such as *Heart Rate Measurement* or *Temperature*, and even proprietary, vendor-specific UUIDs, the meaning of which is assigned by the vendor and depends on the implementation.

Permissions

Permissions are metadata that specify which ATT operations (see "ATT operations" on page 26) can be executed on each particular attribute and with which specific security requirements.

ATT and GATT define the following permissions:

Access Permissions
Similar to file permissions, access permissions determine whether the client can read or write (or both) an *attribute value* (introduced in "Value" on page 55). Each attribute can have one of the following access permissions:

None
The attribute can neither be read nor written by a client.

Readable
The attribute can be read by a client.

Writable
The attribute can be written by a client.

Readable and writable
The attribute can be both read and written by the client.

Encryption
Determines whether a certain level of encryption is required for this attribute to be accessed by the client. (See "Authentication" on page 45, "Security Modes and Procedures" on page 46, and "Security Modes" on page 45 for more information on authentication and encryption.) These are the allowed encryption permissions, as defined by GATT:

No encryption required (Security Mode 1, Level 1)
> The attribute is accessible on a plain-text, non-encrypted connection.

Unauthenticated encryption required (Security Mode 1, Level 2)
> The connection must be encrypted to access this attribute, but the encryption keys do not need to be authenticated (although they can be).

Authenticated encryption required (Security Mode 1, Level 3)
> The connection must be encrypted with an authenticated key to access this attribute.

Authorization
> Determines whether user permission (also known as *authorization*, as discussed in "Security Modes and Procedures" on page 46) is required to access this attribute. An attribute can choose only between requiring or not requiring authorization:

No authorization required
> Access to this attribute does not require authorization.

Authorization required
> Access to this attribute requires authorization.

All permissions are independent from each other and can be freely combined by the server, which stores them in a per-attribute basis.

Value

The attribute value holds the actual data content of the attribute. There are no restrictions on the type of data it can contain (you can imagine it as a non-typed buffer that can be cast to whatever the actual type is, based on the attribute type), although its maximum length is limited to 512 bytes by the specification.

As discussed in "Attribute and Data Hierarchy" on page 56, depending on the attribute type, the value can hold additional information about attributes themselves or actual, useful, user-defined application data. This is the part of an attribute that a client can freely access (with the proper permissions permitting) to both read and write. All other entities make up the structure of the attribute and cannot be modified or accessed directly by the client (although the client uses the handle and UUID indirectly in most of the exchanges with the server).

You can think of the whole set of attributes contained in a GATT server as a table (such as Table 4-1), with each row representing a single attribute and each column representing the different parts that actually constitute an attribute.

Table 4-1. Attributes represented as a table

Handle	Type	Permissions	Value	Value length
0x0201	UUID$_1$ (16-bit)	Read only, no security	`0x180A`	2
0x0202	UUID$_2$ (16-bit)	Read only, no security	`0x2A29`	2
0x0215	UUID$_3$ (16-bit)	Read/write, authorization required	"a readable UTF-8 string"	23
0x030C	UUID$_4$ (128-bit)	Write only, no security	`{0xFF, 0xFF, 0x00, 0x00}`	4
0x030D	UUID$_5$ (128-bit)	Read/write, authenticated encryption required	`36.43`	8
0x031A	UUID$_1$ (16-bit)	Read only, no security	`0x1801`	2

In this fictitious GATT server, the attributes it contains are represented as rows of a simple table. This particular GATT server happens to host only six attributes (a rather low number when compared to real-world devices). Note that, as mentioned earlier in this section, the handles of the different attributes do not need to be immediately consecutive, but the ordinal sequence must progress increasingly, as in this example.

The Value column of the table is intended to mirror the high diversity of formats that attribute values can contain in the different GATT-based profiles. The attributes with handles 0x0201, 0x0202, and 0x031A contain 16-bit integers in their respective value fields. The attribute with handle 0x0215 contains a UTF-8 string, 0x030C contains a 4-byte buffer, and 0x030D holds an IEEE-754 64-bit floating point number in its value field.

Attribute and Data Hierarchy

Athough the Bluetooth specification defines attributes in the ATT section, that is as far as ATT goes when it comes to them. ATT operates in attribute terms and relies on all the concepts exposed in "Attributes" on page 53 to provide a series of precise protocol data units (PDUs, commonly known as *packets*) that permit a client to access the attributes on a server.

GATT goes further to establish a strict hierarchy to organize attributes in a reusable and practical manner, allowing the access and retrieval of information between client and server to follow a concise set of rules that together consitute the framework used by all GATT-based profiles.

Figure 4-1 illustrates the data hierarchy introduced by GATT.

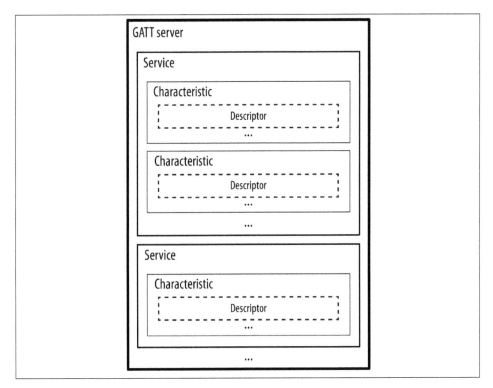

Figure 4-1. GATT data hierarchy

The attributes in a GATT server are grouped into *services*, each of which can contain zero or more *characteristics*. These characteristics, in turn, can include zero or more *descriptors*. This hierarchy is strictly enforced for any device claiming GATT compatibility (essentially, all BLE devices sold), which means that all attributes in a GATT server are included in one of these three categories, with no exceptions. No dangling attributes can live outside of this hierarchy, as exchanging data between BLE devices depends on it.

For most types of data in the GATT hierarchy, it is important to differentiate between their *definition* (the whole group of attributes that make it up) and the *declaration*. The declaration is a single attribute that is always placed first (in increasing handle order) within the definition and that introduces most of the metadata about the data that follows. All declarations have read-only permissions with no security required, because they cannot contain sensitive data. They are only structural attributes that allow the client to find out and discover the layout and nature of the attributes on the server.

Services

GATT services group conceptually related attributes in one common section of the attribute information set in the GATT server. The specification refers to all the attributes within a single service as the *service definition*. Therefore, a GATT server's attributes are in fact a succession of service definitions, each one starting with a single attribute that marks the beginning of a service (aptly named a *service declaration*.) This attribute's type and value format is strictly specified in GATT, as shown in Table 4-2.

Table 4-2. Service Declaration attribute

Handle	Type	Permissions	Value	Value length
0xNNNN	UUID$_{primary\ service}$ or UUID$_{secondary\ service}$	Read Only	Service UUID	2, 4, or 16 bytes

In the declaration shown in Table 4-2, UUID$_{primary\ service}$ (0x2800) and UUID$_{secondary\ service}$ (0x2801) refer to standard, SIG-assigned UUIDs that are used as the exclusive type to introduce a service. They are naturally 16-bit UUIDs (because they are fundamental ones defined by the specfication).

The difference between primary and secondary services is important to note. A *primary service* is the standard type of GATT service that includes relevant, standard functionality exposed by the GATT server. A *secondary service*, on the other hand, is intended to be included only in other primary services and makes sense only as its modifier, having no real meaning on its own. In practice, secondary services are rarely used.

The value of the service declaration attribute itself contains a UUID (as mentioned in "Value" on page 55, the value of an attribute can be any data type), this time corresponding to the UUID of the actual service that this declaration introduces.

Although the service declaration must always be the first attribute of the service, many others can follow it before the next service declaration, usually in the form of characteristics and descriptors.

Conceptually, you could think of a GATT service as a class in any modern object-oriented language, complete with instantiation, because a service can be instantiated multiple times within a single GATT server (however, this is not a common occurrence and most services would therefore be akin to singletons).

Inside a service definition (that is to say, inside a service), you can add one or more references to another services, using *include definitions*. Include definitions consist of a single attribute (the *include declaration*) that contains all the details required for the client to reference the included service.

Included services can help avoid duplicating data in a GATT server. If a service will be referenced by other services, you can use this mechanism to save memory and simplify the layout of the GATT server. In the previous analogy with classes and objects, you could see include definitions as pointers or references to an existing object instance.

Table 4-3 shows the include declaration attribute with all its fields.

Table 4-3. Include Declaration attribute

Handle	Type	Permissions	Value	Value length
0xNNNN	UUID$_{include}$	Read only	Included service handle, end group handle, Included Service UUID	6, 8, or 20 bytes

Again, the UUID$_{include}$ (0x2802) is a special SIG-assigned UUID used exclusively in include declarations, and the value field in this case contains both the start and end handles of the include service, as well as its UUID.

Characteristics

You can understand characteristics as containers for user data. They always include at least two attributes: the *characteristic declaration* (which provides metadata about the actual user data) and the *characteristic value* (which is a full attribute that contains the user data in its value field).

Additionally, the characteristic value can be followed by descriptors, which further expand on the metadata contained in the characteristic declaration. The declaration, value, and any descriptors together form the *characteristic definition*, which is the bundle of attributes that make up a single characteristic.

Table 4-4 shows the structure of the first two attributes of every single characteristic.

Table 4-4. Characteristic declaration and characteristic value attributes

Handle	Type	Permissions	Value	Value length
0xNNNN	UUID$_{characteristic}$	Read only	Properties, value handle (0xMMMM), characteristic UUID	5, 7, or 19 bytes
0xMMMM	Characteristic UUID	Any	Actual value	Variable

All GATT characteristics are always part of a service, and can therefore always be found enclosed in one.

Characteristic declaration attribute

Once again, the characteristic declaration attribute's type UUID (0x2803) is a standardized, unique UUID used exclusively to denote the beginning of characteristics. As with all other declarations (such as service and include), this attribute has read-only permissions, because clients are allowed only to retrieve its value but in no case modify it.

Table 4-5 lists the different items concatenated within the characteristic declaration's attribute value.

Table 4-5. Characteristic declaration attribute value

Name	Length in bytes	Description
Characteristic Properties	1	A bitfield listing the permitted operations on this characteristic
Characteristic Value Handle	2	The handle of the attribute containing the characteristic value
Characteristic UUID	2, 4, or 16	The UUID for this particular characteristic

These three fields are contained in a characteristic declaration attribute value:

Characteristic Properties

> This 8-bit bitfield, along with the additional two bits in the extended properties descriptor (discussed in "Extended Properties Descriptor" on page 62), contains the operations and procedures that can be used with this characteristic. Each of those 9 properties is encoded as a single bit on the bitfield shown in Table 4-6.

Table 4-6. Characteristic properties

Property	Location	Description
Broadcast	Properties	If set, allows this characteristic value to be placed in advertising packets, using the Service Data AD Type (see "GATT Attribute Data in Advertising Packets")
Read	Properties	If set, allows clients to read this characteristic using any of the ATT read operations listed in "ATT operations"
Write without response	Properties	If set, allows clients to use the Write Command ATT operation on this characteristic (see "ATT operations")
Write	Properties	If set, allows clients to use the Write Request/Response ATT operation on this characteristic (see "ATT operations")
Notify	Properties	If set, allows the server to use the Handle Value Notification ATT operation on this characteristic (see "ATT operations")
Indicate	Properties	If set, allows the server to use the Handle Value Indication/Confirmation ATT operation on this characteristic (see "ATT operations")
Signed Write Command	Properties	If set, allows clients to use the Signed Write Command ATT operation on this characteristic (see "ATT operations")
Queued Write	Extended Properties	If set, allows clients to use the Queued Writes ATT operations on this characteristic (see "ATT operations")
Writable Auxiliaries	Extended Properties	If set, a client can write to the descriptor described in "Characteristic User Description Descriptor"

The client can read those properties to find out which operations it is allowed to perform on the characteristic. This is particularly important for the Notify and Indicate properties, because these operations are initiated by the server (further detailed in "Server-Initiated Updates" on page 72) but require the client to enable them first by using the descriptor described in "Client Characteristic Configuration Descriptor" on page 62.

Characteristic Value Handle

These two bytes contain the attribute handle of the attribute that contains the actual value of the characteristic. Although it's often the case, you should never assume that this handle will be contiguous (i.e., 0xNNNN+1) to the one containing the declaration.

Characteristic UUID

The UUID of the particular characteristic, this can be either a SIG-approved UUID (when making use of the dozens of characteristic types included in the standard profiles) or a 128-bit vendor specific UUID otherwise.

Continuing with the class and object-orientation analogy, characteristics are like individual fields or properties in that class, and a profile is like an application that makes use of one or more classes for a specific need or purpose.

Characteristic value attribute

Finally, the characteristic value attribute contains the actual user data that the client can read from and write to for practical information exchanges. The type for this attribute is always the same UUID found in the characteristic's declaration value field (as shown in "Characteristic declaration attribute" on page 59). So, characteristic value attributes no longer have types of *services* or *characteristics*, but rather concrete, specific UUIDs that can refer to a sensor's reading or a keypress on a keyboard.

The value of a characteristic value attribute can contain any type of data imaginable, from temperatures in celsius to key scan codes to display strings to speeds in miles per hour—anything that can be usefully transmitted over two BLE devices can fill in the contents of that value.

Characteristic Descriptors

GATT characteristic descriptors (commonly called simply *descriptors*) are mostly used to provide the client with *metadata* (additional information about the characteristic and its value). They are always placed within the characteristic definition and after the characteristic value attribute. Descriptors are always made of a single attribute, the *characteristic descriptor declaration*, whose UUID is always the descriptor type and whose value contains whatever is defined by that particular descriptor type.

You can find two types of descriptors in the different GATT characteristics:

GATT-defined descriptors

These are the fundamental, widely used descriptor types that simply add meta information about the characteristic. The following sections describe the most common ones.

Profile or vendor-defined descriptors

> Regardless of whether a profile is specified and published by the SIG or by a particular vendor, these descriptors can contain all types of data, including additional information regarding the characteristic value, such as the encoding used to acquire the value from a sensor or any other particulars that complement the reading itself.

The following sections describe some of the most commonly used descriptors defined by GATT.

Extended Properties Descriptor

This descriptor, when present, simply contains the two additional property bits, covered in "Characteristic declaration attribute" on page 59 and listed in Table 4-6.

Characteristic User Description Descriptor

As the name implies, this descriptor contains a user-readable description for the characteristic within which it is placed. This is a UTF-8 string that could read, for example, "Temperature in the living room."

Client Characteristic Configuration Descriptor

This descriptor type (often abbreviated CCCD) is without a doubt the most important and commonly used, and it is essential for the operation of most of the profiles and use cases. Its function is simple: it acts as a switch, enabling or disabling server-initiated updates (covered in more detail in "Server-Initiated Updates" on page 72), but only for the characteristic in which it finds itself enclosed.

Why Provide a Notification Switch?

As discussed earlier, a client knows nothing in advance about a server's attributes, so it will need to perform discovery to find out which services, characteristics, and descriptors are present on the server. The server sends server-initiated updates (handle value notifications and handle value indications, introduced in "ATT operations" on page 26) asynchronously whenever a characteristic's value changes, formatted in a packet that only contains an attribute handle and an array of bytes with the value.

If the client has not yet discovered all characteristic and descriptor handles on the server, it might not yet be able to associate the data received in those notifications and indications with a particular type, rendering those over-the-air transactions useless. Furthermore, even when a client can identify all handles with their corresponding services and characteristics, there might be times (perhaps because the application is not visible, perhaps because the client uses only one of many services and characteristics in the server) when the client simply does not wish to receive updates. That is where CCCDs

---

come in, allowing for fine-grained enabling and disabling of notifications and indications for all characteristics that support them.

A CCCD's value is nothing more than a two-bit bitfield, with one bit corresponding to notifications and the other to indications. A client can set and clear those bits at any time, and the server will check them every time the characteristic that encloses them has changed value and might be susceptible to an update over the air.

Every time a client wants to enable notifications or indications for a particular characteristic that supports them, it simply uses a Write Request ATT packet to set the corresponding bit to 1. The server will then reply with a Write Response and start sending the appropriate packets whenever it wants to alert the client of a change in value.

Additionally, CCCDs have two special properties that separate them from other attributes:

Their values are unique per connection
    In *multi-connection* scenarios, in which a central is connected to multiple peripherals and also acting as a GATT server, each peripheral will receive its own copy of the CCCD's value when reading it with ATT.

Their values are preserved across connections with bonded devices
    "Attribute Caching" on page 66 discusses attribute caching in more detail, but that concerns only attribute *handles*. *Values* are typically not stored per-device and the GATT server can reset them between connections. This is not the case with CCCDs among bonded devices: the last value written by a client to a CCCD on the server is guaranteed to be restored upon reconnection, regardless of the time elapsed between connections.

Many protocol stacks have special mechanisms to deal with CCCDs, both from a client's and a server's perspective, because they are so critical to correct operation and to guarantee timely data updates between peers.

Characteristic presentation format descriptor

When present, this descriptor type contains the actual format of the enclosing characteristic value in its seven-byte attribute value. The list of formats available include Booleans, strings, integers, floating-point, and even generic untyped buffers.

Example Service

This section presents an example of a particular service found in many commercial products today. The Heart Rate Service (HRS) exposes the user's heart rate to a monitoring device.

Figure 4-2 illustrates an instance of the HRS on a fictitious server. This would not be the only service contained in the server, so you can see this as a partial slice of the complete set of attributes that a client could access.

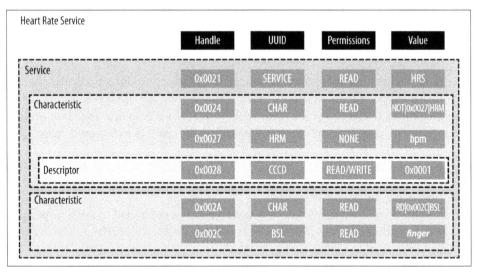

Figure 4-2. GATT Heart Rate Service

Here is a handle-by-handle description of the HRS service illustrated in Figure 4-2:

Handle 0x0021

> This attribute contains the service declaration (see "Services" on page 58) for the Heart Rate Service. These are the service declaration attribute's fields:

> *UUID*

> > The UUID is the standard 16-bit UUID for a primary service declaration, $UUID_{primary\ service}$ (0x2800).

> *Value*

> > The value is the 16-bit UUID for the Heart Rate Service, assigned by the SIG (0x180D).

Handle 0x0024

> This attribute contains the characteristic declaration (see "Characteristic declaration attribute" on page 59) for the Heart Rate Measurement characteristic. These are the characteristic declaration attribute's fields:

> *UUID*

> > The UUID is the standard 16-bit UUID for a characteristic declaration, $UUID_{characteristic}$ (0x2803).

Value

> The characteristic properties for this characteristic are notify only, the characteristic value handle is `0x0027`, and the characteristic value UUID is the UUID for Heart Rate Measurement (`0x2A37`).

Handle `0x0027`

> This attribute contains the characteristic value (see "Characteristic value attribute" on page 61), in this case the heart rate measurement itself. These are the characteristic value attribute's fields:

UUID

> The same UUID present in the last two bytes of the characteristic declaration's attribute value.

Permissions

> This attribute's value is neither readable nor writable: a client can obtain its value only through notifications sent by the server.

Value

> The actual heart rate measurement (fictitiously represented in beats per minute for clarity).

Handle `0x0028`

> This attribute contains a CCCD (a key descriptor described in "Client Characteristic Configuration Descriptor" on page 62). These are the CCCD attribute's fields:

UUID

> The UUID for any CCCD is always the standard 16-bit $UUID_{CCCD}$ (`0x2902`).

Permissions

> A CCCD must always be readable and writable. The security level required to execute these operations is defined by the profile or application.

Value

> As already established, the CCCD's value is a bitfield, in this case `0x0001`, denoting that notifications are enabled for this particular HRM characteristic.

Handle `0x002A`

> This attribute contains another characteristic declaration (see "Characteristic declaration attribute" on page 59), this time for Body Sensor Location characteristic. These are the characteristic declaration attribute's fields:

UUID

> The UUID is the standard 16-bit UUID for a characteristic declaration, $UUID_{characteristic}$ (`0x2803`).

Value

> The characteristic properties for this characteristic are read only, the characteristic value handle is 0x002C, and the characteristic value UUID is the UUID for the Body Sensor Location (0x2A38).

Handle 0x002C

This attribute contains the characteristic value (see "Characteristic value attribute" on page 61), in this case the body sensor location. These are the characteristic value attribute's fields:

UUID

> The same UUID present in the last two bytes of the characteristics definition's attribute value.

Permissions

> This attribute's value is read only: a client can only check where the sensor is located, but not modify its location (that is up to the server).

Value

> The actual body sensor location (fictitiously represented as "finger" for clarity).

Several tools exist to discover and display the different services on a server in a format similar to Figure 4-2, which can be useful during application development. Chapter 6 has more information on these tools.

Advanced Attribute Concepts

This section introduces additional concepts related to working with attributes that are also worth mentioning because their understanding is frequently required for many types of BLE applications.

Attribute Caching

"Attributes" on page 53 discusses how attribute handles allow a client to individually address all available attributes on a server. Discovering the list of available handles and the contents of their respective attributes can be a time-consuming (and power-consuming) process detailed further in "Service and Characteristic Discovery" on page 69. But for now, this section describes what a client can do to avoid performing the discovery procedure every time it reconnects to a server, and under what circumstances it can do so.

Servers typically tend to maintain a stable set of attributes, and their basic structure does not, in most cases, change over the lifetime of a server device. But the implementation imposes no rigid restrictions in this regard, and a server is indeed free to completely overhaul its attributes and even replace them with a radically new set at any time (through a firmware update or perhaps with the installation of applications on the server device, for example). Certain rules and constraints are therefore required, so that a client can rely on the validity of previously discovered handles without risk of them having changed on the server and thus no longer being valid.

As a general rule, the specification recommends that clients cache (i.e., store for subsequent transactions and even connections) the handles of the attributes they are interested in. Attribute values, especially in the cases where they correspond to actual user data, are highly volatile, so it usually makes little sense to store them locally in the client for future use.

The specification provides the Service Changed characteristic (discussed in more detail in "GATT Service" on page 73) for servers to communicate to the client any potential changes in the contents of its attribute information. This is an optional characteristic, so its mere presence on the server already acts as an alert regarding the actual possibility of structural attribute changes.

Clients can ascertain if the result of discovery can be cached for future use by observing the following conditions:

No Service Changed characteristic present on the server
> Clients can freely and permanently cache all handles found with no restrictions. The server guarantees they will not change during the lifetime of the device.

Service Changed characteristic present on the server
> In this case, a client needs to subscribe to the server-initiated updates by writing into the corresponding CCCD enclosed within the Service Changed characteristic (see "Characteristic Descriptors" on page 61). This will allow the server to alert the client of any structural changes. If the client and the server are bonded as described in "Security Modes and Procedures" on page 46, the client can cache attribute handles across connections and expect them to remain identical. If the devices are not bonded, the client will need to perform discovery every time it reconnects to the server.

"GATT Service" on page 73 provides more details about using the Service Changed characteristic.

GATT Attribute Data in Advertising Packets

Although GATT primarily relies on established connections between a central and a peripheral (as described in "Roles" on page 36), it is also possible to include portions of the attribute information hosted by the server inside advertising packets, rendering one or more server attributes available to any observer or central while scanning.

Table 3-3 discusses the Service Data AD Type, but that section does not describe the format it uses to enclose server attributes inside an advertising packet. The Core

Specification Supplement in the Specification Adopted Documents page (*http://bit.ly/ 1gg4UhU*) specifies the fields that the GATT server must insert in the payload of an advertising packet to make a particular service's data available to scanners.

As shown in Table 4-7, to be able to broadcast service data, a GATT server must include two different fields in the advertising packet's Service Data section.

Table 4-7. Service Data AD Type

Field	Length in bytes	Description
UUID	2, 4, or 16	The actual UUID identifying the data
Service Data	Variable	The data associated with the service identified by the UUID

The contents of the Service Data field can correspond to the complete or partial value of a particular characteristic or descriptor within the corresponding service. It is up to each profile specification to define which, because only the profile has sufficient knowledge about the data to decide which pieces of information are the most relevant to be broadcasted.

Features

GATT features are strictly defined procedures that allow GATT-based data exchanges to take place. They are all based on the different operations that ATT provides (introduced in "ATT operations" on page 26).

To a certain extent, most of the features listed in this chapter are exposed in one way or another in most GATT APIs. GATT server APIs add the ability to populate the actual server with attributes, but that is heavily implementation dependant and beyond the scope of this chapter.

Exchange MTU

This succinct two-packet procedure allows each ATT peer to let the other end know about the maximum transmission unit (MTU, or effectively maximum packet length) it can hold in its buffers and can therefore accept.

This procedure is used only whenever either the client or the server (or both) can handle MTUs longer than the default ATT_MTU of 23 bytes (see "Logical Link Control and Adaptation Protocol (L2CAP)" on page 25) and wants to inform the other end that it can send packets longer than the default values that the specification requires. L2CAP will then fragment these bigger packets into small Link Layers packets and recombine them from small Link Layers packets.

Service and Characteristic Discovery

As mentioned elsewhere in this chapter, the client has no knowledge about the attributes that might be present in a GATT server when it first connects to it. It is therefore essential for the client to begin by performing a series of packet exchanges to determine the amount, location, and nature of all the attributes that might be of interest. As discussed in "Attribute Caching" on page 66, procedures in this category can, in some cases, subsequently be skipped.

For *primary service discovery*, GATT offers the two following options:

Discover all primary services
> Using this feature, clients can retrieve a *full list* of all primary services (regardless of service UUIDs) from the remote server. This is commonly used when the client supports more than one service and therefore wants to find out about the full service support on the server side. Because the client can specify a handle range when issuing the required request, it must set `0x0001-0xFFFF` as the handle range to implement this feature, covering the full attribute range of the server.

Discover primary service by service UUID
> Whenever the client knows which service it is looking for (usually because it supports only that single service itself), it can simply look for *all instances* of a particular service using this feature, also with the requirement of setting the handle range to `0x0001-0xFFFF`.

Each of these procedures yield handle ranges that refer to the attributes that belong to a single service. The *discover all primary services* feature also obtains the individual service UUIDs.

When the client has already found services on the server, it can proceed to perform *relationship discovery* (the discovery of any included services) with the following feature:

Find included services
> This allows a client to query the server about any included services within a service. The handle range provided in such a query refers to the boundaries of an existing service, previously obtained using service discovery. As with service discovery, the client also receives a set of handle ranges, along with UUIDs when applicable.

In terms of *characteristic discovery*, GATT offers the following options:

Discover all characteristics of a service
> Once a client has obtained the handle range for a service it might be interested in, it can then proceed to retrieve a *full list* of its characteristics. The only input is the handle range, and in exchange, the server returns both the handle and the value of all characteristic declaration attributes enclosed within that service (see "Characteristic declaration attribute" on page 59).

Discover characteristics by UUID
> This procedure is identical to the previous one, except the client discards all responses that do not match the particular characteristic UUID it targets.

Once the boundaries (in terms of handles) of a target characteristic have been established, the client can go on to *characteristic descriptor discovery*:

Discover all characteristic descriptors
> Now in possession of a set of handle ranges and UUIDs for some or all of the characteristics in a service, the client can use this feature to retrieve all the descriptors within a particular characteristic. The server replies with a list of UUID and handle pairs for the different descriptor declarations (see "Characteristic Descriptors" on page 61).

All the features in this section can be performed over open, unsecured connections, because discovery is allowed for all clients without any restrictions.

Reading Characteristics and Descriptors

To obtain the current value of a characteristic value or a descriptor, the client has the following choices:

Read characteristic value or descriptor
> This feature can be used to simply read the contents of a characteristic value or a descriptor by using its handle. Only the first ATT_MTU-1 bytes of the contents can be read, because that is the maximum number of bytes that can fit in the response packet (1 byte is reserved for the ATT operation code).

Read long characteristic value or descriptor
> If the value is too long to be read with the previous feature, this feature includes an offset along with the handle in the request, so that the characteristic value or the descriptor contents can be read in successive chunks. Multiple request/response pairs might be required, depending on the length of the attribute value being read.

Additionally, and only applicable for characteristic values, these features are available:

Read characteristic value using characteristic UUID
> Whenever a client does not know the specific handles for the characteristics it might be interested in, it can read the values of all the characteristics of a specific type. The client simply provides a handle range and a UUID and receives an array of values of characteristics enclosed in that range.

Read mutiple characteristic values
> Conversely, if a client already has the handles for a set of characteristics it wants to obtain the value from, it can then send a request with this set of handles and subsequently receive the values for all corresponding characteristics.

Reading characteristics and descriptors is subject to the security permissions and the server can deny permission if the connection's security level does not match the established requirements (see "Security" on page 72).

Writing Characteristics and Descriptors

To write to the value of a characteristic value or a descriptor, the client has the following choices:

Write characteristic value or descriptor
> This feature is used to write to a characteristic value or descriptor. The client provides a handle and the contents of the value (up to ATT_MTU-3 bytes, because the handle and the ATT operation code are included in the packet with the data) and the server will acknowledge the write operation with a response.

Write long characteristic value or descriptor
> Similar to the *read long characteristic value or descriptor* feature, this permits a client to write more than ATT_MTU-3 bytes of data into a server's characteristic value or descriptor. It works by queueing several *prepare write* operations, each of which includes an offset and the data itself, and then finally writing them all atomically with an *execute write* operation.

Additionally, and only applicable for characteristic values, these features are available:

Write without response
> This feature is the converse equivalent of notifications (detailed in "Server-Initiated Updates" on page 72) and uses Write Command packets. Write Commands are unacknowledged packets that include a handle and a value, and they can be sent at any time and in any amount without any flow control mechanism kicking in (except of course for the native Link Layer flow control, since all traffic is subject to it). The server is free to silently discard them if it cannot process them or if the attribute permissions prevent it from accepting it. The client will never know, but that is by mutual agreement. The only way for the client to find out whether the value was written is to read it after the fact.

Reliable writes
> Similar to the *read multiple characteristic values* feature, when a client wants to queue write operations to multiple characteristic values, it issues a final packet to commit the pending write operations and execute them.

Writing characteristics and descriptors is subject to the security permissions, and the server can deny permission if the connection's security level does not match the established requirements (see "Security" on page 72).

Server-Initiated Updates

Server-initiated updates are the only asynchronous (i.e., not as a response to a client's request) packets that can flow from the server to the client. These updates send timely alerts of changes in a characteristic value without the client having to regularly poll for them, saving both power and bandwidth.

There are two types of server-initiated updates:

Characteristic Value Notification
> *Notifications* are packets that include the handle of a characteristic value attribute along with its current value. The client receives them and can choose to act upon them, but it sends no acknowledgement back to the server to confirm reception. Along with *write without response*, this is the only other packet that does not comply with the standard request/response flow control mechanism in ATT, as the server can send any number of these notifications at any time. This feature uses the *handle value notification* (HVN) ATT packet.

Characteristic Value Indication
> *Indications*, on the other hand, follow the same handle/value format but require an explicit acknowledgment from the client in the form of a *confirmation*. Note that although the server cannot send further indications (even for different characteristics) until it receives confirmation from the client (because this flows in the opposite direction than the usual request/response pairs), an outstanding confirmation does not affect potential requests that the client might send in the meantime. This feature uses the *handle value indication* (HVI) and *handle value confirmation* (HVC) ATT packets.

The client must enable both types of server-initiated updates by writing to the corresponding CCCD before the server can start sending them. "Client Characteristic Configuration Descriptor" on page 62 describes this process extensively.

Security

"Security Modes and Procedures" on page 46 discusses how the GAP authentication procedure can be used to transition from one security mode to a higher, more secure one by using the different means available within the Security Manager and GAP. GATT transactions can act as triggers of such an authentication procedure. As discussed in "Permissions" on page 54, each attribute within a GATT server has fine-grained, independent permissions for both reading and writing, and these permissions are enforced at the ATT level.

Generally speaking, attributes that are *declarations* require no special security to be accessed. This is true for both service and characteristic declarations, but **not** for descriptor declarations, which contain the relevant data directly in them, rather than in a separate attribute. This is done so that clients that have not yet paired or bonded with a server can at least perform basic service and characteristic discovery, without having to resort to performing security procedures. The attribute layout and data hierarchy of a server is not considered to be sensitive information and is therefore freely available to all clients.

When accessing a characteristic value or a descriptor declaration (also called *service request*), however, a client can receive an *error response* ATT packet (see "ATT operations" on page 26), indicating that the connection's current security level is not high enough for the request to be executed. The following two error codes are commonly used for this purpose and placed in the error response packet:

Insufficient Authentication
> Denotes that the link is not encrypted and that the server does not have a long-term key (LTK, first introduced in "Security Keys" on page 31) available to encrypt the link, or that the link is indeed encrypted, but the LTK used to perform the encryption procedure is not authenticated (generated with man-in-the-middle protection; see "Authentication" on page 45) while the permissions required authenticated encryption.

Insufficient Encryption
> Denotes that the link is not encrypted but a suitable LTK is available.

GAP and GATT roles are not linked in any way and can be mixed and matched freely, but security procedures are always initiated by the GAP central (see "Security Manager (SM)" on page 28). Therefore, depending on which peer is acting as a central and which as a peripheral, it can be up to either the GATT client or the GATT server to initiate the pairing, bonding, or encryption procedure in order to raise the security level of the connection. Once the security level matches the one required by the attribute's permissions, the client can send the request again to be executed on the server.

GATT Service

Just as GAP has its own SIG-specified service that is mandatory for all devices (described extensively in "GAP Service" on page 50), GATT also has its own service (containing up to one characteristic) that must be included in all GATT servers. The optional *service changed* characteristic (introduced briefly in "Attribute Caching" on page 66), cannot be read or written, and its value is communicated to the client only through *characteristic value indications*.

As shown in Table 4-8, the value consists only of a handle range, which delimits a particular area of attributes in the server. This is the area that has been affected by structural changes and needs to be rediscovered by the client. The client will have to perform service and characteristic discovery in that area, because the attributes it can have cached might no longer be valid.

Table 4-8. Service changed characteristic value

Handle	Type	Permissions	Value	Value length
0xNNNN	UUID*service changed*	None	Affected handle range	4

A client must enable indications on the corresponding CCCD for this characteristic before doing anything else, so that it can become aware of any changes on the server's attribute structure.

If the server suffers a structural change in attribute layout, it will then immediately send a handle value indication to the client and wait for the corresponding confirmation. In this way, it can be sure that the client understands that cached attribute handles in that range might no longer be valid. If the attributes change outside the lifetime of a connection with a bonded device, the server will send the indication right after the connection is set up, so that the client has a chance to rediscover the affected area.

CHAPTER 5

Hardware Platforms

Most commercial use cases for BLE-enabled products involve *peripherals*, rather than the *central* devices (phones, tablets, or personal computers) you would design products to interact with. As such, this chapter introduces some specific development platforms for designing and prototyping BLE peripherals.

The discussion in this chapter assumes a basic familiarity with embedded system design (Chapter 10), and the primary goal is to point product designers to inexpensive and readily available platforms that might be appropriate for their products.

nRF51822-EK (Nordic Semiconductors)

Nordic Semiconductors has been involved in low-power wireless solutions for years and, as a board member on the Bluetooth SIG, has helped define and shape the core BLE standard since its inception. Widely known in the wireless market for its popular, general-purpose radio-frequency (RF) silicon solutions, it was one of the first companies to get affordable BLE peripheral-mode silicon to market (the nRF8001). Its newest nRF51 family represents a complete redesign from many of their previous single-chip RF products, combining a radio with a modern 32-bit ARM microprocessor in a single chip.

Technical Specifications

Nordic's nRF51 series is a highly integrated system-on-chip, combining a BLE-compatible radio and a modern ARM processor in a single low-cost package with the following characteristics:

- ARM Cortex-M0 core running at 16 MHz
- 128 or 256 KB flash memory (between 80 and 90 KB is required for the S110 BLE stack)

- 16 KB SRAM (8 KB available to the application with the S110 BLE stack)

Being an entirely *flash-based* device is a key distinguishing factor of the nRF51822 for product designers. This means that the BLE stack is written into modifiable flash memory and can be updated as the core spec evolves, without necessarily requiring a new silicon revision.

The down side of this choice is that it adds some additional per-device manufacturing costs, compared to a ROM-based solution. But given the fast-paced development of the Bluetooth Core Specification, the gamble might pay off in the long run, because it gives Nordic the potential to get to market with support for the latest spec faster than other silicon vendors who have opted for lower-cost ROM-based chips.

SoftDevice Architecture

In order to implement BLE support on its chips (which can also be used for non-BLE standards, such as ANT+ and proprietary 2.4GHz protocols), Nordic makes use of something it calls a *SoftDevice* (SD). The SoftDevice is essentially a black box that sits in the bottom part of flash memory and implements features such as the BLE stack and peripheral role support. The user (application) code sits at a higher address in flash memory and makes calls to this lower-level SoftDevice as appropriate.

Most BLE products use the S110 SoftDevice, which is a peripheral-only solution. The device architecture also includes a S120 SoftDevice that supports the *central* role, but because that use-case is less common for BLE, the discussion in this section will focus on the S110.

Nordic's SoftDevice design approach has its benefits and drawbacks. On the positive side, having a separate, verified, reliable BLE stack in the form of a SoftDevice allows firmware engineers to focus more energy on their application-level code. They can deal with a smaller API and higher-level concepts such as GAP (Chapter 3) and GATT (Chapter 4), leaving the lowest-level details to the SoftDevice (security implementation, message validation, etc.).

The SoftDevice also allows firmware developers to avoid dealing with radio configuration themselves, which can be a significant part of the firmware development process in most RF products. Turning these details into a black box protects firmware engineers from making low-level mistakes and can significantly simplify the product validation process, because the low-level BLE code is guaranteed to function according to the Bluetooth Core Specification.

Another advantage of the SoftDevice approach is it allows one hardware design to support multiple radio protocols or use cases, including the ability to design a custom protocol, which might lower product costs in companies with multiple similar in-house products based on a single PCB design.

Finally, the SoftDevice's architecture is nonintrusive to application developers. Because it runs independently and the application does not need to link against any set of libraries, both SD and application updates can be performed separately with no dependencies, much in the way one can update a Linux kernel or user-space libraries independently without worrying about the other side being affected by the update.

On the down side, the SoftDevice requires system resources that will not be available for your own application's use. The S110 SoftDevice allocates the bottom 80 KB of flash and 8 KB of SRAM, leaving you with 176 KB flash and 8 KB SRAM for your own application (assuming the 256 KB version of the nRF51822 is being used).

The SoftDevice design also introduces latency and architectural limitations, as higher-level code needs to make calls down to the SoftDevice from your higher-level code, which happens through sofware interrupts on the ARM core.

As with any big engineering task, the numerous benefits the SoftDevice brings to the table require some sacrifices, mainly around timing and hard real-time requirements.

Working with the nRF51822-EK

If you are interested in evaluating the nRF51822, the best platform to begin with is Nordic's nRF51822-EK (*http://bit.ly/1iDC4de*). As shown in Figure 5-1, this evaluation kit includes two development boards: the PCA10001 (shown on the left) and the PCA10000 (shown on the right).

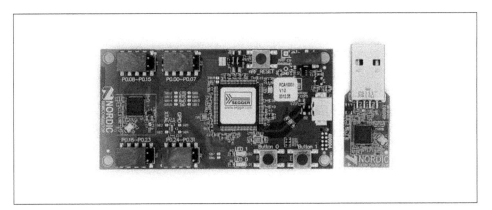

Figure 5-1. Nordic Semiconductor's nRF51822-EK

The PCA10000 is a small USB dongle used primarily for debugging, either by simulating a central device via Nordic's Master Control Panel or as a sniffer to push data out to Wireshark (both of which are discussed in Chapter 7). But it's also a full-featured development board on its own, with an integrated Segger J-Link that allows you to program and debug the firmware using a variety of development tools.

The larger PCA10001 is the main development board in the kit. It breaks out all of the pins available on the nRF51822, enabling you to connect I2C or SPI sensors or peripherals, talk to other devices over UART, etc. This board also includes a J-Link on board to program and debug the MCU, as well as some additional circuitry to measure power consumption and make prototyping and debugging easier. The UART ouput is optionally directed out over USB, for example, so you can easily visualize debug messages or send simple commands back to the MCU. It also features a CR2032 battery holder to allow the board to be powered by a small battery.

Examples and Toolchains

Nordic provides numerous examples (after registering your kit, as discussed in the note following this paragraph) based on this development kit, making it an easy choice if you just want to get started with something you know will work. Most of the demo code uses Keil's uVision (*http://www.keil.com/uvision/*) (with the IAR toolchain as a second option), which is freely available from ARM for noncommercial use on projects smaller than 32 KB (which should actually cover a broad range of projects, because the SoftDevice is not included in the 32 KB limit, only the application code).

 You will need to create a MyPages account on Nordic Semiconductor's website (*http://www.nordicsemi.com/*) and register the serial number for your nRF51822-EK before you can access the latest demo code and development tools for this chipset.

Some support materials and sample projects are also available for GNU-based tools, although GNU (and Eclipse) support is not as extensive as the support for Keil uVision. However, it is worth noting that developing full applications on GNU/Linux and Mac OS X is currently possible without having to resort to Windows.

To provide an open source option, we've also included a basic code base in the GitHub repository for this book (*https://github.com/microbuilder/nRF51822_GNU*). This code is based on the freely available GNU toolchain, including makefiles, startup code, and some basic tools to program the flash using the onboard J-Link debuggers.

CC2541DK-MINI (Texas Instruments)

Another company with a long history in the low-power RF field, Texas Instruments has designed a number of BLE system-on-chips (SoCs) targetting the peripheral market, and was the first to market with a peripheral BLE solution.

The CC2541 combines the following features:

- 8051 core with a 2.4 GHz radio
- 128 or 256 KB user-programmable flash memory
- 8 KB SRAM

One of the biggest advantages TI has over some of its rivals is that its BLE stack is feature complete, essentially covering the entire 4.0 version of the Bluetooth Core Specification. Some vendors have chosen not to implement certain infrequently used optional features, whereas TI has made an intentional effort to reach a bit further in its functionality coverage.

The CC2541 is also pin-compatible with the CC2540, which adds USB support to the SoC. This potentially extends the lifespan of any design efforts around this chip family with minimal additional design effort, since you can easily move your designs over to peripherals that are connected to desktop or laptop PCs via USB connectivity.

TI takes design and testing seriously, and its RF chip families have a long design lineage. It makes reliable chips using well-designed radios and provides significant design resources around those chips, for both hardware designers and firmware engineers. This can be an important consideration for small companies without a lot of in-house RF or embedded firmware design expertise.

One big weakness of the CC2541 is the relatively dated 8051 core driving its SoC, which not only requires an expensive commercial compiler and IDE to use (IAR Embedded Workbench (*http://www.iar.com/Products/IAR-Embedded-Workbench/8051/*)), but also feels quite long in the tooth compared to more modern ARM Cortex-M cores making their way into many other SoCs. This will likely change in the near future, as SoC vendors feel greater pressure to move to newer cores, and TI is no doubt well aware of the trend. It will be interesting to see how the company responds moving forward.

Among the many development kits available for TI's CC254x family, the low-cost CC2541DK-MINI (Figure 5-2) platform should enable peripheral designers to thoroughly evaluate the development platform and SoC.

This kit includes all of the hardware you'll need to start working with the CC2541, including a hardware debugger, a USB dongle that can act as a BLE master device on your PC, and a key fob development board that can run your custom BLE code.

One attractive element of this development kit is that it closely resembles a real product, with an injection-molded enclosure and two large, physical buttons that provide real feedback during the initial development and debugging process. It's easy to overlook these kinds of details in the early stages of product development, but they can help give an idea of the real-world performance of a BLE device, even if some specifics (such as operating range) will of course vary from one device design to another.

Figure 5-2. Texas Instruments' CC2541MINI-DK development kit

Simply holding a low-power device in your hand will often attentuate the signal, providing a very different experience than a raw PCB sitting unobstructed on a desk a few feet away from your laptop. It's also easier to get a battery-powered device in an enclosure off the work bench and do something with it in the real world, such as attaching it to something or someone else. It's a small detail, but it can save you both time and potential surprises later if you didn't get your development kit off your desk early on in the design process.

For further information on this development kit, including ordering details, see the CC2541 product page (*http://www.ti.com/tool/cc2541dk-mini*) on the Texas Instruments website.

Other Hardware Platforms and Modules

If you'd rather not build your own RF devices and circuit boards from scratch, *modules* provide an alternative approach. One of the key advantages of modules is that they typically come precertified as intentional emitters by the various regulatory bodies, such as the FCC or CE/ETSI, and are likely to pass any test programs set up by the various protocol bodies, such as the Bluetooth SIG. FCC or CE certification can easily cost $10,000 per product, making modules an attractive option for products produced in relatively low volume.

Another advantage of using a module is that the RF design is done for you, whereas custom RF hardware design requires specialized knowledge, tools, and testing. Properly designing an antenna or RF front end on a product is a nontrivial task, and poor design can significantly affect the operating range and efficiency of your products. Module makers generally solve this problem for you by providing a properly designed and tuned RF front end and antenna or a common connector that allows you to easily add an external antenna to your product without having to worry about impedance matching on the transmission line (the metal trace that transmits energy to and from the radio).

As another advantage, some modules (e.g., the Bluegiga modules or Laird modules discussed later in this section) come with high-level development scripting languages that can significantly reduce development time and avoid some of the difficulties of working with low-level programming enviroments such as Keil's uVision for Nordic's nRF51822-EK ("nRF51822-EK (Nordic Semiconductors)" on page 75) or IAR for Texas Instruments' CC2541MINI-DK ("CC2541DK-MINI (Texas Instruments)" on page 78). They only require a text editor for code development.

Of course, the benefits of using modules come at a cost. The per-unit price of modules is significantly higher than designing your own hardware using individual integrated circuits (such as the nRF51822 or CC2541). Module makers spread out the design, verification, and certification costs across multiple products and allow low-volume products to hit a price point they wouldn't be able to on their own, but at some point (probably upward of 10,000 units), designing and certifying your own hardware might be more cost effective.

The rest of this section describes three BLE modules available at the time of this writing.

Laird's BL600 Module

Laird's BL600 module (*http://bit.ly/1jyfMLR*) is based on Nordic Semiconductor's nRF51822 ("nRF51822-EK (Nordic Semiconductors)" on page 75). In addition to all of the raw functionality included in the nR51822, these modules add an event-driven smartBASIC programming language that allows you to easily create basic applications without having to learn or invest in expensive commercial IDEs and compilers, or having to program in low-level languages like C or C++.

You're free to program the modules directly using standard C code and Nordic's SDK and toolkit for the nRF51822, but the smartBASIC option might be useful for simple use cases in which you just need to add a wireless link to your product with a minimum of development effort and without having to learn a new stack and technology in depth.

This module has regulatory certifications for CE/ETSI (Europe), FCC, Industry Canada, Japan, and the NCC (Taiwan), as well as Bluetooth SIG Qualification, and is also available for purchase from many major component resellers online.

Bluegiga's BLE112/BLE113 Modules

Bluegiga's BLE112 (*http://bit.ly/1hGyY5H*) and BLE113 (*http://bit.ly/1hyshbS*) modules are based on the CC2540/CC2541 from Texas Instruments ("CC2541DK-MINI (Texas Instruments)" on page 78). They include support for BGScript, which allows you to program certain types of applications using simple XML files. Bluegiga also provides a

C API to work with these modules using an external MCU, talking to the modules over UART.

The main difference between the BLE112 and the BLE113 is that the newer BLE113 module has slightly lower power consumption and adds a hardware I2C port, which can be used to talk to a wide variety of low-cost sensors (temperature sensors, accelerometers, gyroscopes, pressure sensors, etc.)

These modules have the regulatory certifications for CE/ETSI (Europe), FCC, Industry Canada, Japan, and South Korea and are available for purchase from most major component resellers (Digikey, Mouser, Farnell, etc.).

RFDuino

The RFDuino (*http://bit.ly/1hyxuR1*) is a small BLE module that allows you to use the popular Arduino IDE and development platform to create BLE devices. If you're already familiar with Arduino, the RFDuino is an excellent entry point to experiment with BLE, because it can significantly lower the learning curve involved in getting an initial prototype up and running.

These modules have the regulatory certifications for CE/ETSI (Europe), FCC, and Industry Canada and are available from several component resellers, including Mouser and Arrow.

CHAPTER 6

Debugging Tools

This chapter introduces useful several debugging and development tools for working with Bluetooth Low Energy. It includes hardware tools, such as wireless protocol analyzers or *sniffers* (which sniff traffic over the air, displaying the captured data in a UI to analyze later), and tools to interact directly with BLE peripherals during the debug process.

To keep things accessible to small startups or engineers and developers who are just getting started with BLE, this chapter focuses on inexpensive tools, rather than high-end products that can run thousands or tens of thousands of dollars.

PCA10000 USB Dongle and the Master Control Panel

The PCA10000 is a USB dongle included in Nordic Semiconductor's nRF51822-EK ("Working with the nRF51822-EK" on page 77), a low-cost evaluation kit for the nRF51822 system-on-chip (SoC). While this kit is designed for embedded hardware engineers who are designing their own BLE peripherals, it might be worth purchasing the kit even if you are only developing mobile applications, because it includes a number of extremely useful debug tools for a relatively modest price tag.

One of these tools is the Master Control Panel (MCP), a Windows-based utility that turns the PCA10000 USB dongle into something that can simulate a BLE central device. It features an easy-to-use interface that allows you to see any data available on BLE peripherals in range or send data back to any peripheral you are connected to. This is particularly useful on Windows 7, which doesn't include native support for Bluetooth Low Energy (BLE support was introduced in Windows 8, but that operating system doesn't include a comparable application for testing and debugging).

Nordic also provides a Master Control Panel application for Android that includes some of the same functionality without any additional hardware requirements, though the standalone tool using the PCA10000 supports a much larger command set at the time of this writing.

If you are writing an application for an existing peripheral, you can also use the MCP to quickly reverse-engineer BLE accessories, displaying their individual data structures and config settings, and then access them in your mobile application using the discovered service and characteristic UUIDs.

The MCP talks to the PCA10000 using a special firmware image included with the tool's installer. Using Nordic's nRFGo Studio (also available on Nordic's website after registering your nRF51822-EK, per the note in "Examples and Toolchains" on page 78), you can update the USB dongle to use this firmware image.

Once the PCA10000 has been updated with the appropriate firmware, you can open the MCP and interact with your peripheral via an easy-to-navigate UI, which allows you to perform almost any functionality of a normal central device. This includes bonding, opening or closing a connection, reading and writing to GATT characteristics, etc.

Figure 6-1 shows the results of a single peripheral advertising itself.

Once you connect to the periperhal and send a service discovery request, you can see the list of services and characteristics available on the device (as shown in Figure 6-2), at which point you can read or write to them the same way you could with any regular BLE central device.

You can update values by selecting an appropriate characteristic, modifying the value in the Value textbox, and then clicking "Send update." You can also retrieve the latest value of any characteristic by selecting it and clicking the Read button, which can be useful with characteristics for which notification or indication have not yet been enabled.

Figure 6-1. Master Control Panel displaying advertising data

The MCP is an extremely valuable tool early in the hardware development process, when you might not have a mobile application for your BLE peripheral to talk to yet. The Master Control Panel can simulate almost any action that your application will perform, including validating communication for both incoming and outgoing data.

The MCP also includes a set of C# libraries that can be used to automate any of the functionality it provides, allowing application developers to create desktop or command-line applications that have access to a simple but complete set of central APIs. This can be extremely useful for automated regression testing or production tests.

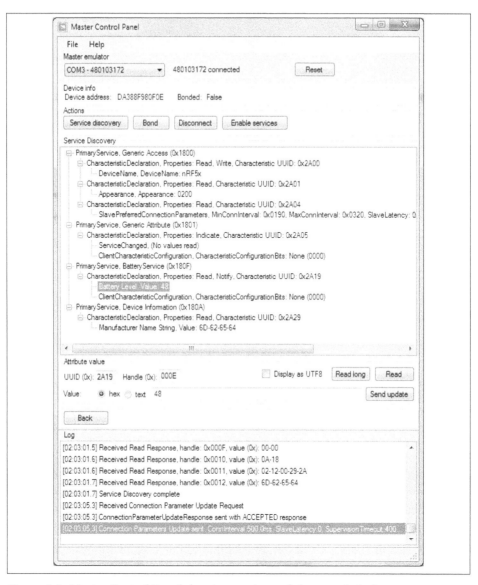

Figure 6-2. Master Control Panel showing service and characteristic data

PCA10000 USB Dongle and Wireshark

The Master Control Panel ("PCA10000 USB Dongle and the Master Control Panel" on page 83) is likely the easiest way to interact with BLE peripherals, but some use cases require lower-level access to the BLE data. For those cases, Nordic also provides a custom firmware image and tools for the PCA10000 or PCA10001 (both of which are included

---

in their nRF51822-EK development kit) that can *sniff* traffic from a single peripheral device and push it into Wireshark (*http://www.wireshark.org/*).

Wireshark is a mature and powerful open source data-capture and analysis tool that allows you to easily visualize data down to the packet and byte level. Nordic's Wireshark plugin (shown in Figure 6-3), which is available after registering your nRF51822-EK kit at Nordic's support site (*http://www.nordicsemi.com/*), takes data captured via the nRF51822-EK boards and adds useful descriptions to help make sense of that raw data.

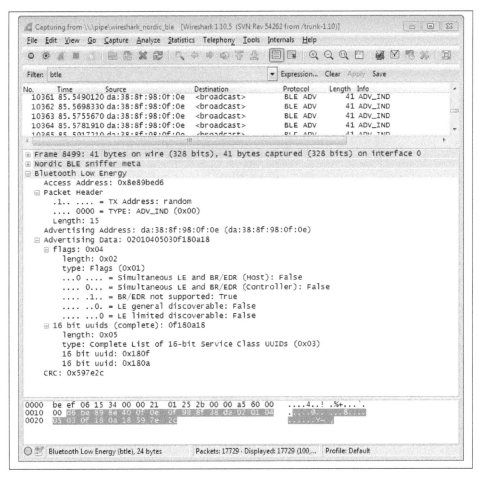

Figure 6-3. Nordic Semiconductor's PCA10000 breakout with the Wireshark plugin

If you are interested only in designing applications that talk to existing BLE peripherals, you will rarely need to go down to this level. But it can be extremely useful for hardware designers or firmware engineers working on their own peripheral designs and code or trying to debug a particular latency or throughput issue.

CC2540 USB Dongle and SmartRF Sniffer

As part of the development ecosystem around their CC254x family of ICs, Texas Instruments designed the CC2540EMK-USB (*http://www.ti.com/tool/cc2540emk-usb*) (Figure 6-4), a low-cost CC2540-based USB dongle that can be used with their free SmartRF software (*http://www.ti.com/tool/packet-sniffer*) (Figure 6-5) to convert the board into a BLE sniffer. This combination allows you to see all BLE data going out over the air around you at the lowest possible level.

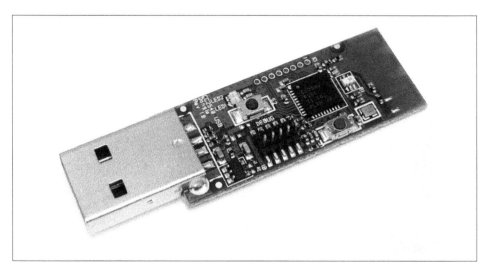

Figure 6-4. CC2540EMK-USB

This performs a function similar to the PCA10000/Wireshark combination ("PCA10000 USB Dongle and Wireshark" on page 86), but it provides a different UI that might be easier to work with in certain situations. The kits might be easier to source in certain regions.

SmartRF-to-Wireshark Converter

If you prefer Wireshark ("PCA10000 USB Dongle and Wireshark" on page 86) as a data analysis tool and have access to the CC2540 USB dongle ("CC2540 USB Dongle and SmartRF Sniffer" on page 88), you will be happy to find out about smartRFtoPcap (*https://github.com/mikeryan/smartRFtoPcap*), a freely available tool that converts saved SmartRF data into a file format that WireShark can understand.

You won't be able to stream live data into Wireshark the way you can with the PCA10000 and Nordic's Wireshark plugin, but having the option to convert previously captured files might still be useful, since Wireshark includes numerous utilities to filter and search inside your logged data.

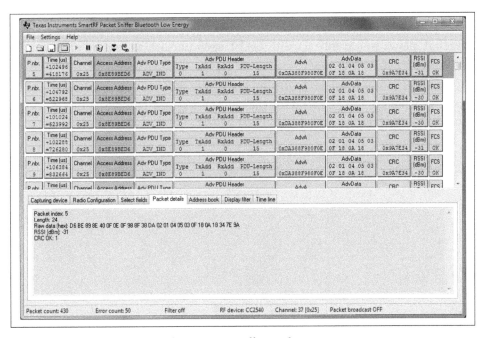

Figure 6-5. Texas Instruments' SmartRF Sniffer application

Bluez hcitool and gatttool

If you are using a Linux workstation, you can take advantage of two useful utilities in the Bluez Bluetooth Stack, `hcitool` and `gatttool`, which allow you to interact with BLE devices from the command line.

If you don't have access to a dedicated Linux workstation, Bluez also runs fine on inexpensive Linux devices, such as the Raspberry Pi or BeagleBone Black, which turns them into extremely useful and portable BLE debug tools

`hcitool` allows you to scan for BLE peripherals in range, connect to them, or optionally simulate a BLE device using any supported BLE 4.0 USB dongle. To scan for BLE devices in range, you can issue the following command (assuming our USB dongle is enumerated as `hci0`):

```
sudo hcitool -i hci0 lescan
```

Once you have the device's address (retrieved via the previous scan command), you can connect to the peripheral using the following command (which assumes a peripheral address of 6C:60:B3:6E:7C:B1):

```
sudo hcitool lecc 6C:60:B3:6E:7C:B1
```

gatttool allows you to interact with GATT services, such as reading or writing to characteristics on your device.

Being able to do this from the command line means that you can easily script certain repetitive actions or test cases and run the same tests across multiple hardware devices consistently and reliably.

CHAPTER 7

Application Design Tools

This chapter introduces useful tools for mobile application developers, including code templates available for common mobile operating systems, inexpensive hardware platforms that can be used to simulate physical BLE devices (useful if you don't have your own hardware available at the early stages of app development), as well as some purely software-based tools that can help debug and develop your application.

Bluetooth Application Accelerator

Bluetooth Low Energy is heavily concentrated on the mobile computing space, specifically mobile phones and tablets, with some adoption on traditional desktop computer and laptops—especially in the *human interface device* (HID) category, mostly for mice and keyboards. In recognition of this, the Bluetooth SIG introduced the Bluetooth Application Accelerator initiative to make adopting Bluetooth Low Energy much easier for application designers.

The Application Accelerator is essentially a set of software templates that make it relatively painless to create basic BLE applications on iOS, Android, and Windows RT 8.1 platforms. It also includes simple example code that shows how to connect and work with GATT services and characteristics.

Some knowledge of the specific platform is required, of course, but the templates should make it relatively easy to get up to speed on the BLE-specific calls required to interact with the target peripheral devices. The program requires registration, but no fees. For more details, see the Bluetooth Developer Portal (*http://bit.ly/1kQPQuv*).

SensorTag

While most people would associate Texas Instruments (TI) with hardware design, as part of its BLE development ecosystem, TI has released an interesting and low-cost BLE development tool called SensorTag (Figure 7-1) that is extremely useful to mobile application developers.

Figure 7-1. Texas Instruments' SensorTag BLE device

Mobile applications are obviously only one half of the picture for most BLE use cases, and you generally need hardware for your application to interact with. But hardware might not be available in the early stages of the development process, or if you are simply interested in learning to work with the BLE APIs on a mobile platform, you might not know where to start finding a device to talk to.

The SensorTag solves this problem by providing an impressive collection of sensors in a relatively inexpensive and easy-to-order device, with example apps for both iOS and Android that show how to interact with those sensors.

The battery-powered Bluetooth Low Energy SensorTag platform includes the following sensors:

- Temperature sensor
- Humidity sensor
- Pressure sensor
- Accelerometer
- Gyroscope
- Magnetometer

This combination of sensors opens up a number of unique opportunities for data collection and rich user interaction. The accelerometer and magnetometer can be combined to collect reasonably accurate three-axis rotational data, the pressure sensor can measure changes in altitude as the device moves up or down, and so on.

Most importantly, the SensorTag platform allows application developpers to talk to real hardware and get real sensor data with a minimum of fuss and without having to learn how to design peripherals themselves or make progress on the app while they wait for working hardware from other specialists working on the project.

For more details, or to order a device, visit TI's SensorTag product page (*http://www.ti.com/tool/cc2541dk-sensor*).

LightBlue for iOS

Many mobile application developers might not be interested in designing their own BLE hardware, but instead want to talk to existing BLE peripherals that are already available on the market.

While some of the debugging tools discussed in Chapter 6 will no doubt be useful to app designers, anyone with a recent iOS device has access to an extremely useful tool from Punch Through Design called LightBlue. This free application, which is available in the App Store (*http://bit.ly/1hq3m9j*), can be used to reverse engineer or even emulate any BLE peripheral from an iPhone or an iPad.

You can use LightBlue to interact with services and characteristics a device exposes—for example, reading or writing individual values. You can also use it to *capture* the unique signature of an existing BLE peripheral once and then play that signature back for development purposes, essentially emulating the device that you might not have access to day to day.

If you know what the device should look like, but real hardware isn't available, you can also create a profile to imitate your peripheral and then create a device that will appear exactly the same as the hardware once it is available, as shown in Figure 7-2.

A version of LightBlue is also available for OS X in Apple's App Store, although the iOS version offers more functionality at the time of this writing.

Application developers can also use LightBlue when developing a BLE peripheral but before the counterpart iOS or Android application has been completed. You can simulate the over-the-air interactions between central and peripheral, allowing you to debug the peripheral's firmware without yet having a custom application developed on the central.

Figure 7-2. LightBlue showing a BLE device simulated on another instance of LightBlue

nRF Master Control Panel for Android

If you are running Android 4.3 or higher on a device that supports Bluetooth Low Energy, the Android version of Nordic's Master Control Panel allows you to debug, reverse-engineer, or interact with existing BLE hardware in an easy-to-use UI.

You can download this free tool from Google's Play Store (*http://bit.ly/1n13gYS*) and find the UUID for any service or characteristic present on nearby BLE peripherals, subscribe to notifications sent by the characteristics, or write values back to the peripheral, similar to the way LightBlue (see "LightBlue for iOS" on page 93) and the PC-based Master Control Panel application (see "PCA10000 USB Dongle and the Master Control Panel" on page 83) work.

Figure 7-3 shows the results of a Heart Rate Monitor service, where the Body Sensor Location and Heart Rate Measurement are both visible.

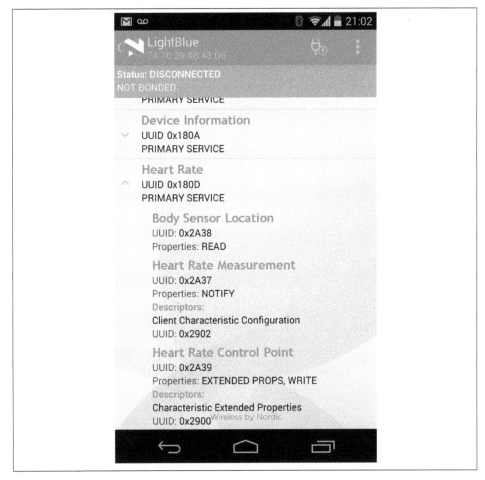

Figure 7-3. nRF Master Control Panel showing a Heart Rate Monitor service

Further information about the Master Control Panel app and other Android applications from Nordic Semiconductors is available on its support pages (*http://bit.ly/ 1kQGE9s*).

CHAPTER 8

Android Programming

One of the main benefits of using a wireless standard like Bluetooth Low Energy is its support for a wide variety of smartphones and tablets. This opens up a new dimension of interaction to embedded hardware projects, which can now be designed with cheap hardware and rich interfaces.

Beyond the interface, you can also use the phone as a gateway to the larger Internet or in conjunction with other apps or APIs to create custom mashups with the embedded hardware you've created. This enables a whole new class of inexpensive devices that still offer rich functionality.

This chapter provides a basic overview of the hardware, software, and processes required to implement Bluetooth Low Energy on the Android operating system.

Getting Started

The example Android project developed in this chapter interfaces with the low-cost SensorTag device ("SensorTag" on page 91) manufactured by Texas Instruments (TI). The SensorTag offers many sensors and is a great example of a complex sensor device that can provide lots of information to be processed and visualized.

Because the GUI side of Android can get a bit complex and is generally beyond the scope of this book, this chapter focuses on getting you to the point where you can extract data out of the SensorTag and receive it via Bluetooth Low Energy. At that point, many other available resources can demonstrate ways to present data.

Get the Hardware

For hardware, you'll need an Android device running Android version 4.3 or later. While Android began supporting BLE version 4.3, we recommend a device running at least version 4.4, which includes an updated and more stable version of the BLE protocol stack. You'll also need to make sure the hardware supports Bluetooth Low Energy. This example project uses a Google Nexus 7 running Android 4.4.

 To confirm whether your device supports BLE, see Blueooth's Smart Devices List (*http://bit.ly/1iDPs0M*).

You'll also need to pick up a TI SensorTag, which will serve as the peripheral device in this project. See "SensorTag" on page 91 for more information about the SensorTag device.

Get the Software

For this project, you'll need three main pieces of software:

Eclipse Android Development Tools (ADT)
    Available at the Android Developer site (*http://bit.ly/PYcBmE*)

Bluetooth Application Accelerator
    Available at the Bluetooth SIG website (*http://bit.ly/1kQPQuv*)

TI SensorTag Android app source code
    Available at TI's website (*http://bit.ly/1kR0aTd*)

The most important and most time-consuming part of the setup is installing the Android Development Tools, which also requires understanding a bit about how to use the Eclipse IDE. Your best bet is to go to the Android Developer site (*http://bit.ly/PYcBmE*) for detailed instructions on setting up your working environment. You'll need to download the latest SDKs and any updates to the Android Developer Tools.

Configure the Hardware

The Android device needs a bit of configuration before it can be used as a development device. First, you'll need to enable Developer mode, if it's not already enabled. Go to the Settings menu, scroll down to the bottom, and select About. On the About screen (shown in Figure 8-1), tap "Build number" seven times in rapid succession to turn on Developer mode.

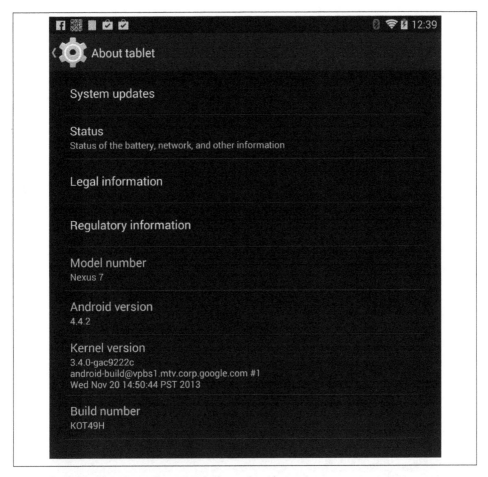

Figure 8-1. Enabling Developer mode from the About screen

You should now see a new "Developer options" item in the Settings menu. Now, you'll need to enable the "USB debugging" and "Stay awake" options, as shown in Figure 8-2.

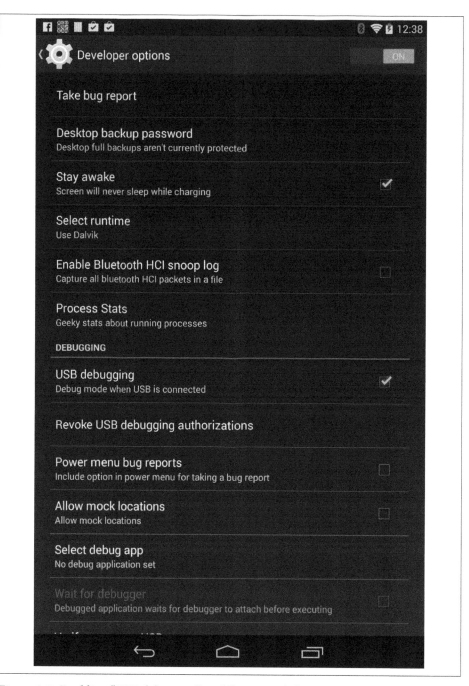

Figure 8-2. Enabling "USB debugging" and "Stay awake" options on the "Developer options" screen

Finally, go to the Settings menu, select Storage→Options→"USB computer connection," and enable the Camera option (as shown in Figure 8-3)to allow you to transfer files.

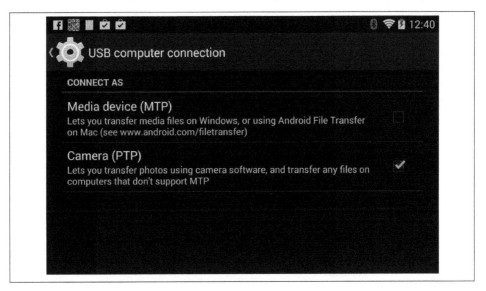

Figure 8-3. Configuring the Android as a camera to transfer files

This might sound counter-intuitive, but the device needs to be in Camera mode for the development tools to work.

Start a New Project

To start a new application, you'll first need to import the Bluetooth Application Accelerator's Android files into the project. Open Eclipse ADT, select File/Import, navigate to the directory that contains the Application Accelerator, and select the Android/BLEDemo folder. Press OK to import the project into the workspace.

At this point, you might want to look through the Application Accelerator files a bit to see how the code is laid out. If it looks foreign to you, don't worry. Rather than spending too much time inside the Application Accelerator, you're going to move the main class files for the BLE library into your own project.

Now it's time to create your own Android project. In Eclipse, go to File→New→Android Application Project. For the Application Name, use BleSensorTag. For Minimum Required SDK, specify Android 4.3. This is the minimum version of Android that supports Bluetooth Low Energy. For Target SDK and Compile With, use the latest version of Android the device supports. Click Next for the rest of the windows to accept the defaults. Finally, the project wizard should create a new Android project for you called BleSensorTag.

As shown in Figure 8-4 (which shows the project directory structure on the left and the main code window in the right), an Android project contains many folders and files, but you'll be working in only a few. The main sourcecode file is located in the */src* directory.

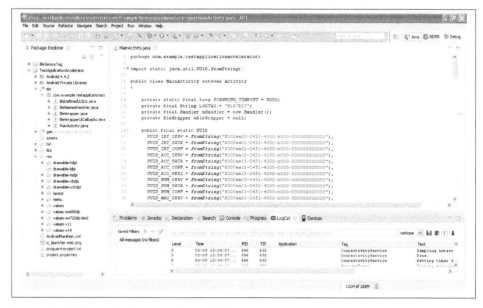

Figure 8-4. The main Java window in Eclipse

Each project also includes a manifest, called *AndroidManifest.xml*, that details essential information about the application to the Android system. Android requires this information before it can run any code on the system. Beyond that, you'll also be working in the */res folder*, which contains XML files that handle the layout and menus that you will be using.

 Android programming is a huge topic in itself, far beyond the scope of this chapter, which focuses only on the context of integrating Bluetooth Low Energy into an Android app. For more complex GUI applications, we recommend consulting other Android programming texts.

Before you begin any actual coding, you'll need to enable BLUETOOTH and BLUETOOTH_AD MIN permissions inside the Android manifest. Double-click the *AndroidManifest.xml* file and select the Permissions tab (shown in Figure 8-5). Choose "Add…"→"Uses Permission" and type android.permission.BLUETOOTH in the Name field. Do the same to add another "Uses Permission," this time giving it the name android.permission.BLUE

TOOTH_ADMIN. These two entries request permission from the user to access the named services when the application is installed.

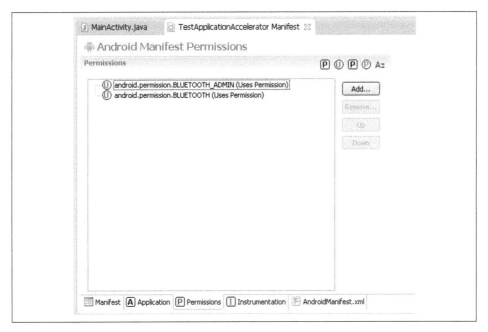

Figure 8-5. The Android Manifest Permissions window

Now, you need to move class files from the Bluetooth Application Accelerator into your project. That will make the classes and methods available to you, many of which make BLE programming much easier. Select and copy the following files:

- *BleWrapper.java*
- *BleWrapperUiCallbacks.java*
- *BleNamesResolver.java*
- *BleDefinedUUIDs.java*

Paste the files into the */src* directory of the BleSensorTag project (or drag and drop them into that directory). Once you've added these files, the */src* directory should look like Figure 8-6.

Figure 8-6. The Android /src directory

`BleWrapper` is the main part of the Application Accelerator library issued by the Bluetooth SIG. It's a simplified wrapper into the Android Bluetooth Low Energy library that makes accessing and using the library much simpler, because it handles a lot of the complicated processing.

This takes care of the prep work for the project. Now that you've created the project, configured the manifest, installed the class libraries, you're ready to get into the meat of the Android code.

Initializing the BLE Library

Now it's time to get into the actual implementation of things. The first time through (to emphasize the BLE aspect of the programming without getting too far into the GUI side of things), the example will mainly print out information received over Bluetooth Low Energy. After that, you'll add some GUI features incorporate the BLE data into the presentation side of things.

Before you begin, you'll need to create an instance of the `BleWrapper` library you imported. This is the file from the Bluetooth Application Accelerator library from the Bluetooth SIG:

```
public class MainActivity extends Activity {

    // add this line to instantiate the BLE wrapper
    private BleWrapper mBleWrapper = null;

    @Override
    protected void onCreate(Bundle savedInstanceState) {
        super.onCreate(savedInstanceState);
        setContentView(R.layout.activity_main);
    }
```

The line added in this file will hold the instance of the `BleWrapper` class, which will also contain the methods you'll be using to access the underlying Android BLE library. In your constructor, you'll also be adding various callbacks that you'll need to handle events from the Android BLE library.

In the previous code snippet, notice an automatically generated function called `onCreate()`. This function is called when the application first starts and you'll use it to intialize things that need to be initialized on startup and only once. This includes members that don't need to be reinitialized if the device goes to sleep or loses context and then comes back.

The Android documentation actually outlines an "activity lifecycle," which describes the method as an app goes through different stages of its "life." For example, when an app is started, three separate lifecycle methods get called: `onCreate()`, `onStart()`, and `onResume()`. Initializing variables inside an app class occurs in `onCreate()` when they're initialized only once, and in `onResume()` when they need to be initialized again if the device goes to sleep or loses context.

The documentation provides more information on this activity lifecycle, but it's an important concept to know, because you'll be doing part of the init routines in on `Create()` and part of them in `onResume()`. You'll also need to handle cleanup when you shut down, which will be in `onPause()` or `onStop()`.

Inside the `onCreate()` method, you'll see some boilerplate code generated by the project wizard. Beneath that, you'll be adding more code to initialize your `mBleWrapper` object and perform other init tasks:

```
@Override
protected void onCreate(Bundle savedInstanceState)
{
    super.onCreate(savedInstanceState);
    setContentView(R.layout.activity_main);

    mBleWrapper = new BleWrapper(this, new BleWrapperUiCallbacks.Null()
    {
    });

    if (mBleWrapper.checkBleHardwareAvailable() == false)
    {
        Toast.makeText(this, "No BLE-compatible hardware detected",
                    Toast.LENGTH_SHORT).show();
        finish();
    }
}
```

The preceding code allocates a new `BleWrapper` object to the `mBleWrapper` member variable (which previously was null) by instantiating it and calling its constructor. When instantiating the wrapper, you also need to add the BLE callbacks as an argument to the constructor. This is where you add code to handle notification events that occur from the BLE stack.

In software programming, a *callback* is a reference to executable code passed in as an argument to a function, where the function is expected to callback the code after it's done processing something. For now, we won't add any callbacks to the constructor, but we'll revisit this later.

After the constructor, the code performs a check to make sure the BLE hardware is available to the system. If it's not, a popup message window called "Toast" notifies the user about the missing BLE hardware and then shuts down the app.

That completes the code needed to execute when the app starts up in the onCreate() method. Now, to move to the onResume() method, which is where the code goes when it starts up and whenever it wakes from sleep:

```
@Override
protected void onResume() {
    super.onResume();

    // check for Bluetooth enabled on each resume
    if (mBleWrapper.isBtEnabled() == false)
    {
        // Bluetooth is not enabled. Request to user to turn it on
        Intent enableBtIntent = new Intent(BluetoothAdapter.
                                        ACTION_REQUEST_ENABLE);
        startActivity(enableBtIntent);
        finish();
    }

    // init ble wrapper
    mBleWrapper.initialize();
    }
}
```

 Eclipse provides a quick shortcut for any of these functions. Just type the first few letters of the method name and then Ctrl+Spacebar. Choose the function from the autocomplete list that pops up to let Eclipse build the function for you.

Here, the boilerplate code is at the top with super.onResume(). You're going to add your own code beneath that. In this case, you need to check if Bluetooth is enabled each time the device comes back from a different context or from sleep. While some other program is in use, Bluetooth might have been turned off. Not catching that change and just assuming Bluetooth was still on would eventually cause some type of malfunction or exception to the software. To handle this situation, if the program finds that Bluetooth is off, it uses Android intents to send a request to the user to turn Bluetooth on and then exits the app. The next time the app restarts, if Bluetooth is on, it can proceed.

Once the code gets past that last check, it can initialize the `BleWrapper`, which opens up the Bluetooth interface and gets an instance of the Android Bluetooth adapter. Once you have this, you can access the BLE radio and protocol functions.

Finally, you need to handle one last Android app lifecycle method: `onPause()`. This method gets called whenever context is lost, the device goes to sleep, or the app is shut down:

```
@Override
protected void onPause() {
        super.onPause();

        mBleWrapper.diconnect();
        mBleWrapper.close();
}
```

In our example, the Android device will act as a GAP central and a GATT client and the SensorTag device will be a GAP peripheral and a GATT server (see "Roles" on page 36 and "Roles" on page 51).

Once again, the boilerplate code is kept on top. After that, it calls two methods from the wrapper. The first method allows you to "diconnect" (sic) our Android device from the remote device. This actually disconnects from the remote peripheral device and also issues a call to the `uiDeviceDisconnected()` callback method. If you need to handle anything after you disconnect from the peripheral, you'd override the `uiDeviceDiscon nected()` callback. Finally, we close the `BleWrapper`, which closes the local GATT client and central completely.

Connecting to a Remote Device

Now that the initialization and cleanup are out of the way, we can focus on the real meat of the code. The principle tasks associated with BLE you'll need to handle are scanning and connecting to remote devices, communicating data, and performing any necessary management or security tasks.

When initiating a scanning procedure (see "Advertising and Scanning" on page 19), you let the BLE library know that you want to be notified of any advertising packets it receives from remote devices. Whenever a device is found, the `BleWrapper` makes a call to the `uiDeviceFound()` callback function with information about the device. You can use the device information to decide if it's the device you want to connect to.

For scanning, you're going to need to create two buttons: the first to start the scan and the second to stop it. To create the buttons, you'll need to go into the project's /res/ menu directory and edit the main.xml file. If you double-click that file, Eclipse will take you to a GUI interface you can use to add menu items. Any menu item in the *main.xml* file will end up as a button in the Options menu. You're going to create two items, one called Start and one called Stop, which will be used to start and stop the scan, respectively.

In the Android Menu GUI, choose Add→Item and use `action_scan` and `action_stop` for the names, as shown in Figure 8-7. These will be the IDs of the buttons you can use when you write the click handler for the menu buttons. In the Title field of the Menu item, type Scan and Stop, respectively, to supply the text that shows up in the menu item.

Figure 8-7. Adding buttons in the Android ADT menu screen

Once the menu items are created, you're going to need to process the button-click events. By default, the Android boilerplate code will use any items in *main.xml* as the Options menu buttons. The boilerplate does not have a click handler, though, so you'll need to add that in.

In Eclipse, if you type `onOptionsItemSelected` and press Ctrl+Spacebar, it should autocomplete the function, including the `@Override` keyword. You'll be overriding the `onOptionsItemSelected()` method to implement your own handler for click events. When a click occurs on any item in the menu, the `onOptionsItemSelected()` method gets called. Once you're inside the method, you need to have a switch statement to handle the event based on whichever menu item was clicked. The pseudocode will look like this:

```
@Override
public boolean onOptionsItemSelected(MenuItem item) {

        switch (item.getItemId())
        {
        case R.id.action_scan:
```

```
        mBleWrapper.startScanning();
        break;

    case R.id.action_stop:
        mBleWrapper.stopScanning();
        break;

    default:
        break;
        }

        return super.onOptionsItemSelected(item);
    }
```

This code overrides the `OptionsItemSelected()` function to install your own handler. This method is called only when processing events related to the Options menu. In the switch statement, you decide which handler to call based on the item ID of the button that was clicked. This item ID is based on the menu item ID of the buttons you created previously. The `R.id` prefix means that it's an ID from the resource directory of the internal Android project and not part of the wider android namespace.

The two actions in the new handler will either start the scan or stop the scanning procedure, as you can see in the two methods from the `BleWrapper`. In fact, starting and stopping the scanning procedure is probably the easiest part of the whole setup to implement the button functionality.

The next step is the action to take when a remote device is detected. A remote device will be advertising it's there and ready to be connected to. When the program goes into scanning mode, it turns on scanning in the BLE radio and lets the Android system know that it wants to be informed if it get any advertisements. The Android Bluetooth library will inform the program via callback.

To make use of the callback, you're going to revisit the code you wrote in the on `Create()` method. Inside that method, you initialized your `BleWrapper`. Inside the constructor of the `BleWrapper`, it expects you to provide a list of callbacks. You previously left that empty, but now you're going to start implementing the callbacks. The first one will override the `uiDeviceFound()` callback, which will get called by the library whenever a device is found during the scan:

```
mBleWrapper = new BleWrapper(this, new BleWrapperUiCallbacks.Null()
    {
        @Override
        public void uiDeviceFound(final BluetoothDevice device,
                                  final int rssi,
                                  final byte[] record
                                  )
        {
            String msg = "uiDeviceFound: "+device.getName()+", "+rssi+",
                    "+rssi.toString();
```

```
        Toast.makeText(this, msg, Toast.LENGTH_SHORT).show();
        Log.d("DEBUG", "uiDeviceFound: " + msg);
    }
}
```

This code posts a toast message box whenever it finds a device. It's not very practical, but it's an easy way to indicate when a device is found and display information about the device. In a real use case, you'd be outputting any found devices to a device list that people could click to initiate a connection.

The toast message displays two pieces of information from the advertising packet: the device name (see "Advertising Data Format" on page 48) and the RSSI. RSSI (a wireless term that stands for *received signal strength indicator*) is a way to indicate how strong the received signal (in this case the average of all the bits contained in the advertising packet) is. In some cases, the RSSI can be used to give a rough approximation of distance. Another key piece of information that would likely be useful is the device address (which is included with every advertising packet), but the device name alone is fine for now.

The code not only outputs the same information to the toast messagebox, but it also logs it via the Log command, which is an Android library. The Log messages show up in the Eclipse IDE during debug using the logcat tool (shown in Figure 8-8). When you use the Log command, you also specify a tag. The good thing about this is that you can filter on the tag to remove extraneous information.

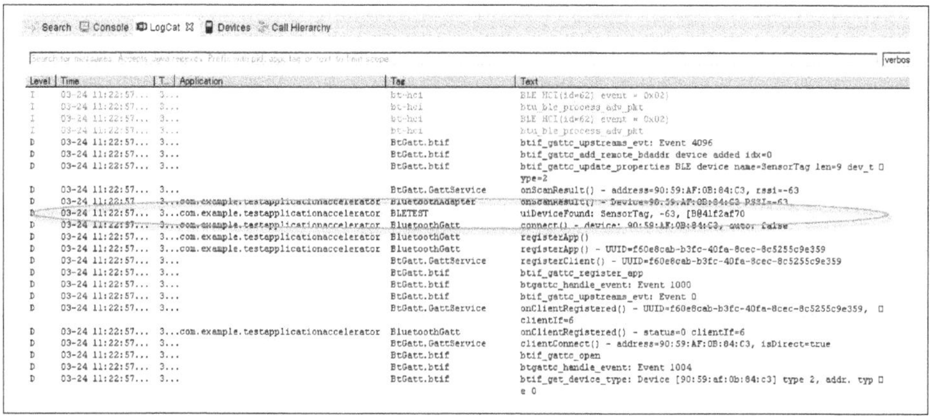

Figure 8-8. The message from the Log command highlighted in the full logcat information dump

By now, the code should have been able to verify that it could detect a remote device sending advertising packets. Once it's able to do that, it needs to be able to connect to it. Normally, you'd save all the detected devices and display them to the user in list form, as explained in "Connection establishment procedures" on page 41. The user would then be able to click on the device she wants to connect to. In order to avoid the GUI complexities, we're going to simplify the use case a bit by making assumptions.

We'll assume we want to connect to a SensorTag device. The following code checks for advertising packets with the device name "SensorTag" and automatically initiates a connection request to any device with that name:

```
@Override
public void uiDeviceFound(final BluetoothDevice device,
                          final int rssi,
                          final byte[] record)
{
    String msg = "uiDeviceFound: "+device.getName()+", "+rssi+", "+ \
        rssi.toString();
    Log.d("DEBUG", "uiDeviceFound: " + msg);
    if (device.getName().equals("SensorTag") == true)
    {
        boolean status;
        status = mBleWrapper.connect(device.getAddress().toString());
        if (status == false)
        {
            Log.d("DEBUG", "uiDeviceFound: Connection problem");
        }
    }
}
```

This automatic connection process avoids going into some Android GUI complexities and lets us focus on the BLE library we're working with.

The previous callback code for uiDeviceFound() has been modified to remove the popup messagebox. After finding a device with a name equal to "SensorTag," the code instructs the BleWrapper to connect to the device and pass in the device's address (introduced in "Bluetooth Device Address" on page 19) in string format. The BleWrapper will use the address to send a unicast connection request to the device. It will then inform you via the uiDeviceConnected() callback if the connection was successful.

You will need to override this callback and add handler code. If the connection was not successful for some reason, the status of the method call will return false. You can use this to log a message to the logcat console for debugging.

Communicating with a Remote Device

Once you're successfully connected to a remote device, the BleWrapper will automatically start service discovery (as described in "Service and Characteristic Discovery" on page 69) for the new device. This means that it will request the new device to list all the services and characteristics from the device and store them in a list.

If the connection to the GATT server and service discovery is successful, another callback is issued. This time, the callback is uiAvailableServices() and one of the arguments is the *service list*, which lists all the Bluetooth GATT services (see "Services" on page 58 for further information on services and characteristics) available on the remote device. In order to communicate with the device, we'll need to access the services and then the characteristics inside the services.

At this time, you could just cycle through the list and print out the services, but they're not in a human-readable format. The services are all listed by 128-bit UUID (see "UUIDs" on page 52). The class library files you copied over earlier form the Bluetooth Application Accelerator included a class called BleNamesResolver. This class has various methods to resolve UUIDs into BLE names. It can resolve both services and characteristics, so this library is quite useful.

The list of known UUIDs is located in the *BleDefinedUUIDs.java* file, which you'll need to add some proprietary service UUIDs to list later on. For now, rather than simply scrolling through the service list and printing out the service UUIDs, you'll be resolving them into human-readable names and printing those out (just to logcat for now). The code is another callback override and goes in the onCreate method and inside the mBleWrapper constructor:

```
@Override
public void uiAvailableServices(BluetoothGatt gatt,
                                BluetoothDevice device,
                                List<BluetoothGattService> services
                                )
{
        for (BluetoothGattService service : services)
        {
                String serviceName = BleNamesResolver.resolveUuid
            (service.getUuid().toString());
                Log.d("DEBUG", serviceName);
        }
}
```

This code cycles through each element in the service list. For each service, it then converts the UUID to a string value and passes it into the BleNamesResolver.resolveUuid() method. This method looks through the list of known UUIDs and, when it finds a match for the UUID, returns the associated human-readable UUID name. It prints that name out to logcat in the Eclipse IDE, as shown in Figure 8-9. It's also possible to dump these directly into a textbox for viewing, but that's slightly more complicated.

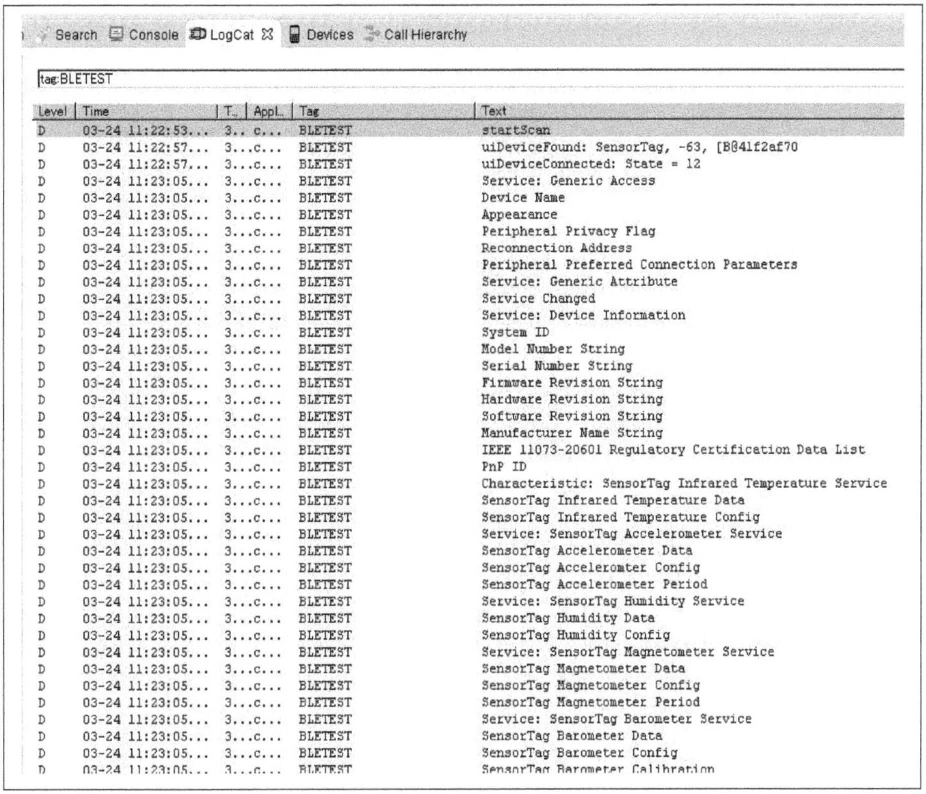

Figure 8-9. Logcat displaying services and characteristics in Eclipse

You'll notice that some UUIDs are unknown. That usually means that vendor specific 128-bit UUIDs are present and you'll have to add them to the UUID list if you want them to be resolved properly. In this case, TI uses some vendor-specific service UUIDs for its device, because it doesn't fit into a standard device profile. There is a huge UUID address space, so you'll likely encounter many vendor specific UUIDs while working with different BLE devices. You'll need to add the TI vendor specific UUIDs to the list and then run the code again.

So far, you've gone most of the way to getting your BLE app up and running. You've detected the remote device, connected to it, and printed the services that it has available. The next step is to read the characteristic values (introduced in "Characteristics" on page 59) associated with its sensors.

The characteristics that belong to the sensors carry the sensor data, so by reading these out, you can get the sensor data. Once you have the data from the sensors, it's just a matter of formatting and processing the data, then presenting it to the user in an appealing way. In this case, we'll focus on getting the data and leave it up to the user to determine how to process and present it.

One thing to be aware of when working with the SensorTag is that it's a mobile device. It was designed to be low power, so the sensors are powered off by default. To read each sensor, you have to write to a characteristic to turn it on. Once the sensor is enabled, you can then read the data from it.

As discussed in "Characteristics" on page 59, all operations related to user data are effectuated through characteristics. To enable a sensor, you're first going to need to find the service that contains the corresponding characteristic and the characteristic itself (see "Service and Characteristic Discovery" on page 69), and then retrieve its value (as described in "Reading Characteristics and Descriptors" on page 70). You then need to modify the characteristic value to one that will enable the sensor and then write it back to the device in the manner described in "Writing Characteristics and Descriptors" on page 71. This is called a *read-modify-write* operation.

You've already connected to the peripheral device and have the `BluetoothGatt` object available to you. To retrieve the service, you're going to use a method inside the `gatt` object called `getService()`. It takes a UUID argument, which means that you'll have to provide the particular service UUID. These are all TI-specific UUIDs, but luckily, TI provides the full list of service and characteristic UUIDs in Java source format in their SensorTag source code.

You'll want to copy and paste the full UUID listing into your source code and make them constants in your app. Here's an example of what it should look like:

```
private static final UUID
    UUID_IRT_SERV = fromString("f000aa00-0451-4000-b000-000000000000"),
    UUID_IRT_DATA = fromString("f000aa01-0451-4000-b000-000000000000"),
    UUID_IRT_CONF = fromString("f000aa02-0451-4000-b000-000000000000"),
    UUID_ACC_SERV = fromString("f000aa10-0451-4000-b000-000000000000"),
    UUID_ACC_DATA = fromString("f000aa11-0451-4000-b000-000000000000"),
    UUID_ACC_CONF = fromString("f000aa12-0451-4000-b000-000000000000"),
    UUID_ACC_PERI = fromString("f000aa13-0451-4000-b000-000000000000");
...
```

The service UUIDs have `SERV` suffix. Otherwise, the other UUIDs are characateristics. Once you have these defined, you can write code to access the particular sensor's service and characteristics.

Communicating with a remote device isn't quite as simple as reading a characteristic and having the method return a value. You actually need to send ATT read and write requests to the device, and the GATT server will then send responses to the requests (these concepts are further explained in "ATT operations" on page 26). Only one request can be handled at one time, and any other request that comes in while a request is being processed gets silently dropped. This can be a source of frustration because it simply looks like a device is not responding.

The correct sequence of operations is to send a request and wait for the appopriate callback. For example, you would send a read request to the remote device to read a particular characteristic. After the device responds, the `BleWrapper` will issue a callback to `uiNewValueForCharacteristic` with the characteristic information. By doing so, you are implementing the *Read chacteristic value* GATT feature explained in "Reading Characteristics and Descriptors" on page 70.

This code requests a read of the accelerometer's configuration characteristic:

```
BluetoothGatt gatt;
BluetoothGattCharacteristic c;

gatt = mBleWrapper.getGatt();
c = gatt.getService(UUID_ACC_SERV).getCharacteristic(UUID_ACC_CONF);
mBleWrapper.requestCharacteristicValue(c);
```

Once the the request is issued, the device will respond with the characteristic's data. In this case, you're going to dump each byte of the characteristic's raw value to `logcat`:

```
@Override
public void uiNewValueForCharacteristic(BluetoothGatt gatt,
                                        BluetoothDevice device,
                                        BluetoothGattService service,
                                        BluetoothGattCharacteristic ch,
                                        String strValue,
                                        int intValue,
                                        byte[] rawValue,
                                        String timestamp)
{
        super.uiNewValueForCharacteristic(  gatt, device, service,
                                        ch, strValue, intValue,
                                        rawValue, timestamp);
        Log.d(LOGTAG, "uiNewValueForCharacteristic");
        for (byte b:rawValue)
        {
                Log.d(LOGTAG, "Val: " + b);
        }
}
```

It's important to remember that each read or write must be requested. There are functions in Android's BLE library to get and set a value for characteristics. These operate only on the locally stored values and not on the remote device. In most cases, any interaction with the remote device will require the use of callbacks.

Before you read any of the sensor's data, you first need to enable them. To do this, you have to write a value to their configuration characteristic (those are proprietary characteristics to enable sensors, *not to be confused with CCCDs*). In most cases, you'd just write a `0x01` to this characteristic to enable them. As mentioned before, you actually send a write request (*Write characteristic value* in "Writing Characteristics and Descriptors" on page 71, with the operation listed in "ATT operations" on page 26) to the peer device and then wait for the callback.

This code enables the accelerometer by writing `0x01` to the accelerometer configuration characteristic on the remote device:

```
BluetoothGattCharacteristic c;

c = gatt.getService(UUID_ACC_SERV).getCharacteristic(UUID_ACC_CONF);
mBleWrapper.writeDataToCharacteristic(c, new byte[] {0x01});
mState = ACC_ENABLE;    // keep state context for callback
```

As before, you need to wait for the callback to know if the write operation was successful. In cases where you have to wait for things to happen, it's best to use a separate thread or a state machine. A separate thread would allow you to block the thread as you wait for callback without holding up the rest of the system. The state machine allows you to remain in the same thread and keep track of the current context for the operation being performed.

For a write operation, Application Accelerator has two useful callbacks:

```
@Override
public void uiSuccessfulWrite(  BluetoothGatt gatt,
                                BluetoothDevice device,
                                BluetoothGattService service,
                                BluetoothGattCharacteristic ch,
                                String description)
{
        BluetoothGattCharacteristic c;

        super.uiSuccessfulWrite(gatt, device, service, ch, description);

        switch (mState)
        {
        case ACC_ENABLE:
                Log.d(LOGTAG, "uiSuccessfulWrite: Successfully enabled
            accelerometer");
                break;
        }
}

@Override
public void uiFailedWrite(  BluetoothGatt gatt,
                            BluetoothDevice device,
                            BluetoothGattService service,
                            BluetoothGattCharacteristic ch,
                            String description)
{
        super.uiFailedWrite(gatt, device, service, ch, description);
```

```
            switch (mState)
            {
            case ACC_ENABLE:
                    Log.d(LOGTAG, "uiFailedWrite: Failed to enable accelerometer");
                    break;
            }
    }
```

To enable all of the sensors, you would basically extend the case of enabling one sensor to all of the sensors. After the write operation to enable the first sensor, you would enable each of the other sensors in the callback:

```
    @Override
    public void uiSuccessfulWrite(  BluetoothGatt gatt,
                                    BluetoothDevice device,
                                    BluetoothGattService service,
                                    BluetoothGattCharacteristic ch,
                                    String description)
    {
        BluetoothGattCharacteristic c;

            super.uiSuccessfulWrite(gatt, device, service, ch, description);

            switch (mState)
            {
            case ACC_ENABLE:
                    Log.d(LOGTAG, "uiSuccessfulWrite: Successfully enabled
                        accelerometer");

                // enable next sensor
                c = gatt.getService(UUID_IRT_SERV).getCharacteristic(UUID_IRT_CONF);
                mBleWrapper.writeDataToCharacteristic(c, new byte[] {0x01});
                mState = IRT_ENABLE;    // keep state context for callback
                    break;

        case IRT_ENABLE:
            Log.d(LOGTAG, "uiSuccessfulWrite: Successfully enabled IR temp sensor");

            // enable next sensor
            c = gatt.getService(UUID_HUM_SERV).getCharacteristic(UUID_HUM_CONF);
            mBleWrapper.writeDataToCharacteristic(c, new byte[] {0x01});
            mState = HUM_ENABLE;        // keep state context for callback
            break;

        case HUM_ENABLE:
            ....
            mState = MAG_ENABLE;
            break;

        ...
            }
    }
```

Once the sensors are enabled, it's possible to read the sensor. To manually read the sensor, you'll issue a read request to the characteristic you want to read and wait for the response in the callback (see "ATT operations" on page 26 and "Reading Characteristics and Descriptors" on page 70 for further details). The read can be triggered by an event, such as the push of a button or a timer that periodically polls the sensor. In this case, a test button created in the Options menu triggers events. That button's onClick method calls the following function to generate a read request:

```
// Start the read request
private void testButton()
{
    BluetoothGatt gatt;
    BluetoothGattCharacteristic c;

    if (!mBleWrapper.isConnected()) {
        return;
    }

    Log.d(LOGTAG, "testButton: Reading acc");
    gatt = mBleWrapper.getGatt();
    c = gatt.getService(UUID_ACC_SERV).getCharacteristic(UUID_ACC_DATA);
    mBleWrapper.requestCharacteristicValue(c);
    mState = ACC_READ;
}

// Get the read response inside this callback
@Override
public void uiNewValueForCharacteristic(BluetoothGatt gatt,
                                        BluetoothDevice device,
                                        BluetoothGattService service,
                                        BluetoothGattCharacteristic ch,
                                        String strValue,
                                        int intValue,
                                        byte[] rawValue,
                                        String timestamp)
{
    super.uiNewValueForCharacteristic( gatt, device, service,
                                       ch, strValue, intValue,
                                       rawValue, timestamp);

    // decode current read operation
    switch (mState)
    {
    case (ACC_READ):
        Log.d(LOGTAG, "uiNewValueForCharacteristic: Accelerometer data:");
        break;
    }

    // dump data byte array
    for (byte b:rawValue)
    {
```

```
        Log.d(LOGTAG, "Val: " + b);
    }
}
```

Once the sensor's data is acquired, a few steps are still needed to process the data. The code for the data processing of each sensor can be found in the SensorTag example Android library or in the SensorTag Online User Guide (*http://bit.ly/1kQHyTq*).

You can copy/paste the processing algorithm for the accelerometer into the previous code snippet, usually in the ACC_READ state part of the decoder. The same goes for the data processing for all the other sensors.

Polling the remote sensor is one way to periodically extract data from it, but it's not power efficient for both devices. The remote device needs to stay on constantly to catch the read requests and the phone needs to stay awake to send the polling requests. A more efficient way to do things is to have the device send notifications (thoroughly described in "Server-Initiated Updates" on page 72) instead. You can set the period for the notifications manually for some of the sensors.

Application Accelerator has a special method for enabling notifications. To enable notifications on Android, you normally have to locally enable the notification for the particular characteristic you are interested in.

Once that's done, you also have to enable notifications on the peer device by writing to the device's client characteristic configuration descriptor (CCCD), as described in "Client Characteristic Configuration Descriptor" on page 62. Luckily, this is abstracted out and both operations can be handled by a single method call.

The following code enables notifications for the accelerometer when the test button is pushed:

```
private void testButton()
{
    BluetoothGatt gatt;
    BluetoothGattCharacteristic c;

    if (!mBleWrapper.isConnected()) {
        return;
    }

    // set notification on characteristic
    Log.d(LOGTAG, "Setting notification");
    gatt = mBleWrapper.getGatt();
    c = gatt.getService(UUID_ACC_SERV).getCharacteristic(UUID_ACC_DATA);
    mBleWrapper.setNotificationForCharacteristic(c, true);
    mState = ACC_NOTIFy_ENB;
}
```

One issue to look out for in the Application Accelerator is it doesn't implement the onDescriptorWrite() callback available in the Android Bluetooth library. The most reliable way to know when notifications have been successfully enabled on the peer device is to get the onDescriptorWrite() callback after the CCCD has been modified and the GATT server has acknowledged the write operation. This example adds the onDescriptorWrite() callback to this implementation of Application Accelerator in the BleWrapper class, where it implements the BluetoothGattCallback code:

```
/* callbacks called for any action on particular Ble Device */
private final BluetoothGattCallback mBleCallback = new BluetoothGattCallback()
{
...

// Added by Akiba
@Override
public void onDescriptorWrite(  BluetoothGatt gatt,
                                BluetoothGattDescriptor descriptor,
                                int status)
{
        String deviceName = gatt.getDevice().getName();
        String serviceName = BleNamesResolver.resolveServiceName( \
        descriptor.getCharacteristic().getService().getUuid().\
        toString().toLowerCase(Locale.getDefault()));
        String charName = BleNamesResolver.resolveCharacteristicName(\
        descriptor.getCharacteristic().getUuid().toString().\
        toLowerCase(Locale.getDefault()));
        String description = "Device: " + deviceName + " Service: " \
        + serviceName + " Characteristic: " + charName;

        // we got response regarding our request to write new value to
        // the characteristic, let's see if it failed or not
        if(status == BluetoothGatt.GATT_SUCCESS) {
                mUiCallback.uiSuccessfulWrite( mBluetoothGatt, mBluetoothDevice,
                                        mBluetoothSelectedService,
                                        descriptor.getCharacteristic(),
                                        description);
        }
        else {
                mUiCallback.uiFailedWrite( mBluetoothGatt, mBluetoothDevice,
                                mBluetoothSelectedService,
                                descriptor.getCharacteristic(),
                                description + " STATUS = " + status);
        }
};

...
}
```

The onDescriptorWrite() method will then call Application Accelerator's uiSuccess fulWrite() method if the notification was written properly.

You can enable notifications for all of the sensor characteristics in the same way you enabled the sensors themselves, using a state machine to sequentially process one after the other. For notifications, keep in mind that, for Android 4.4 (KitKat), only four characteristics can have enabled notifications at one time. This is a limitiation of the current Android BLE library implementation, although it will probably be changed in future versions.

The code in this chapter is mostly presented in fragments, but you can pick up the full source code from the GitHub repository for this book (*http://bit.ly/1qoj8Ed*).

iOS Programming

Apple was an early supporter of Bluetooth 4.0 and, as a result, there is a rich set of APIs and tools to support development of BLE devices and applications using iOS. The iOS device in question is usually an iPhone (iPhone 4S or later), but iOS also supports BLE on all relatively new iPads (iPad 3 or later, or any iPad mini) and the fifth generation iPod Touch devices.

BLE is also supported on later-generation Macs—including iMac (built after late 2012), MacBook Pro (2012 or later), MacBook Air (2011 or newer), and Mac Pro (2013 or later)—but this chapter will focus on iOS, specifically iOS 7 and later.

The types of BLE devices and applications most relevant to iOS programming fall into three main categories:

Peripheral devices with iOS apps
> In this category, BLE peripheral devices and sensors are paired with corresponding iOS apps—for example, a bicycling power and cadence meter that uses the iPhone to display and record data.

iBeacon devices
> The iBeacon is a broadcast-only device that uses BLE advertising (see "Advertising and Scanning" on page 19) to augment navigation indoors, where GPS signals and cell phone towers typically can't penetrate, to provide location services to iOS devices (and also Android devices).

Peripheral devices with Apple Notification Center Service
> Built into iOS, the Apple Notification Center displays alerts and notifications, such as incoming caller ID and updates from news services, on the screen of the iOS device. Using the Apple Notification Center Service (ANCS), iOS devices can use BLE to display notifications on an auxiliary display, such as a BLE-enabled watch.

Developing apps for sensors primarily invloves using the Core Bluetooth framework, while developing apps for iBeacon primarily involves the Core Location framework (ANCS does not require an app). The Core Bluetooth framework (`CoreBluetooth`) is the portion of the iOS API (application programming interface) that deals with BLE functions and devices. Its `CBCentralManager` and `CBPeripheral` classes allow you to work with the centrals and peripherals described in Chapter 3. `CBCentralManager` provides resources for scanning, discovering, and connecting to remote peripherals, while `CBPeripheral` provides resources for working with services and characteristics that are part of the remote peripheral.

This chapter provides practical familiarity with these frameworks by studying code examples that focus on BLE functionality, especially the key classes and methods needed to implement BLE-enabled apps. These examples illustrate the foundations upon which a complete BLE-based app can be built. While the examples are not complete and polished apps, they are fully functional on an iOS device, and understanding how they work is a good start toward writing your own BLE-enabled apps for iOS.

Complete Xcode projects for all examples are available in the Git-Hub repository for this book (*http://bit.ly/1qoj8Ed*). All examples require iOS 7 or later and Xcode 5 or later.

A general discussion of app development for iOS and the Xcode development environment is a broad and complex topic in itself, most of which is beyond the scope of this chapter or book. If you're looking for more information, we recommend *Programming iOS 7* and *iOS 7 Programming Fundamentals* (O'Reilly) by Matt Neuburg. Of course, the most authoritative source is Apple iOS Developer Library, beginning with the Core Bluetooth Programming Guide (*http://bit.ly/1eh8soX*). The examples in this chapter will closely follow the approach recommended in that guide.

Simple Battery-Level Peripheral

This first example programs a central iOS device to look for and connect to a simple remote peripheral. The role of the remote peripheral is played by a Bluegiga BLE112 hardware module (see "Bluegiga's BLE112/BLE113 Modules" on page 81 for more information in this module). The iOS device will act as a GATT client and the peripheral as a GATT server (see "Roles" on page 51).

The Bluegiga BLE112 module evaluation board used here is available from many online suppliers as part of a development kit for the BLE112 module family (part #: DKBLE112). The kit includes a number of prebuilt sensors and inputs for quick evaluation. Alternatively, Jeff Rowberg's BLE112 Bluetooth Low Energy board (*http://bit.ly/OQ3seE*) is much less expensive. This is an open source hardware solution, so all the hardware design details are available from the same source.

The BLE112 module performing the function of the BLE peripheral in this case reads a voltage set by adjusting a small potentiometer (circled in the upper-left corner of Figure 9-1) using the on-board A/D converter and stores that value as the BLE "battery level" characteristic (scaled from 0 to 100%) via the BLE `battery_level` service. The iOS app must then read the "battery level" stored on the BLE peripheral and make it available for use by the app. (For more information on services and characteristics, see Chapter 4).

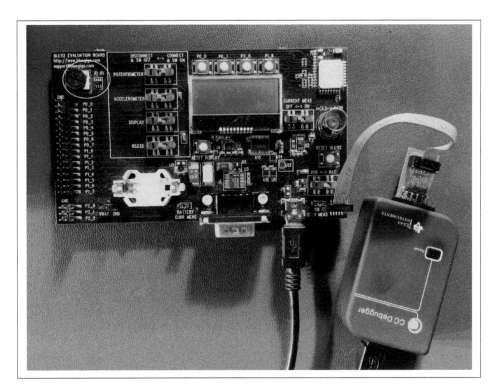

Figure 9-1. Bluegiga BLE112 development breadboard

We are using the term *battery level*, rather than *potentiometer*, because a "battery level" peripheral is one of the predefined services and associated characteristics in BLE. Though there is no requirement to use a predefined profile, doing so is convenient for this discussion.

For this simple application, the key class references within the Core Bluetooth framework include CBCentralManager (an abstraction of the central manager role in BLE on the iOS device) and CBPeripheral (an abstraction of the remote peripheral on the iOS device). The code also includes the supporting classes CBService and CBCharacteristic, because you'll need to know which services and related charateristics are available on the remote peripheral by discovering them (see "Service and Characteristic Discovery" on page 69).

Figure 9-2 shows the relationship between the services and characteristics programmed into the BLE remote peripheral device and the CBPeripheral objects.

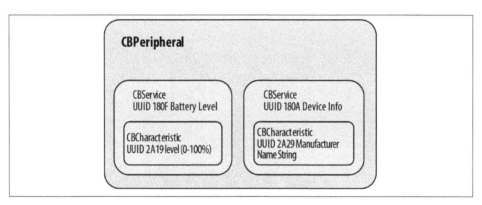

Figure 9-2. Relationship of CBPeripheral, CBServices, and CBCharacteristic objects

For convenience, the code in this example includes a few predefined services and charateristics (e.g., the Battery Service) from the BLE specification, which use 16-bit UUIDs instead of vendor-specific 128-bit UUIDs (see "UUIDs" on page 52).

For more information on SIG-specified services and characteristics, the Bluetooth Developer Portal provides a complete list of all predefined services (*http://bit.ly/1giCbMM*) and adopted characteristics (*http://bit.ly/1ggImgY*).

Scanning for Remote Peripherals

To get started, you'll first need to allocate an instance of the `CBCentralManager` and start the BLE scanning (see "Advertising and Scanning" on page 19) operation for the battery level service:

```
// using the predefined BLE UUIDs for the battery level service
#define BATTERY_LEVEL_SERVICE_UUID 0x180f
#define BATTERY_DEVICE_INFO_SERVICE_UUID 0x180a

// build an array of services UUIDs of interest
NSArray *services = @[[CBUUID UUIDWithString:BATTERY_LEVEL_SERVICE_UUID],
            [CBUUID UUIDWithString:BATTERY_DEVICE_INFO_SERVICE_UUID]];

// instantiate and start Central Manager object
CBCentralManager *centralManager = [[CBCentralManager alloc]
                        initWithDelegate:self queue:nil];

// scan for peripherals with services listed in services UUID array
[centralManager scanForPeripheralsWithServices:services options:nil];

self.centralManager = centralManager;
```

First, you use Core Bluetooth framework to create a list (`NSarray`) of the service UUIDs of interest, named `services` in this case. You will need to convert the UUID strings into CBUUID (*http://bit.ly/1qHI9bB*) objects corresponding to UUIDs using the `UUIDWith String` method.

CBUUID also converts the predefined 16-bit service and characteristic UUIDS into the their fully formed 128-bit equivalents (for more information, see "UUIDs" on page 52). Next, the `CBCentralManager` object is instantiated and the list of UUID objects is provided as the first parameter for its `scanForPeripheralsWithServices` method. In this case, the app scans for just two services, the device info service ($UUID_{16}$=0x180A) and the battery level service ($UUID_{16}$=0x180f). Only peripherals that include the required service UUIDs in their advertising data (see Table 3-3) will be returned as found.

Note that if the `scanForPeripheralsWithSevices` method were given `nil` for the first parameter instead of the list of CBUUIDs of interest (`services`), it would return all BLE peripherals in range. This is generally not a good practice, unless you really want to explore all the possible peripherals with services, because it works the BLE radio hardware more, which decreases battery life.

Connecting to Remote Peripherals

When a peripheral of interest is discovered in the scanning process, the `centralManag er` invokes `didDiscoverPeripheral`, which is where you include the delegate method connectPeripheral. This method returns associated instances of peripheral objects that can be examined later using CBPeripheral methods:

```
- (void)centralManager:(CBCentralManager *)central
  didDiscoverPeripheral:(CBPeripheral *)peripheral
      advertisementData:(NSDictionary *) advertisementData
                   RSSI:(NSNumber *)RSSI
{
    NSString *localName = [advertisementData objectForKey:
                              CBAdvertisementDataLocalNameKey];

    if ([localName length] > 0)
    {
        NSLog(@"Found the battery level monitor: %@", localName);

        // found peripheral so stop scanning
        [self.centralManager stopScan];

        // return peripheral object
        self.batteryLevelPeripheral = peripheral;
        peripheral.delegate = self;

        // connect peripheral to centralManager
        [self.centralManager connectPeripheral:peripheral options:nil];
    }
}
```

It is good practice to stop scanning (using [self.centralManager stopScan]) after the peripheral of interest is found, again because it preserves battery power. Of course, if you need to find more than one peripheral, modify the logic in this example.

Looking Up Services Associated with a Remote Peripheral

After the peripherals are discovered and the peripheral object instantiated, you need to create related service objects and associated characteristics:

```
- (void)centralManager:(CBCentralManager *)central
  didConnectPeripheral:(CBPeripheral *)peripheral
{
    // establish peripheral delegate
    [peripheral setDelegate:self];

    // discover available services
    [peripheral discoverServices:nil];

    // log results
    self.connected = [NSString stringWithFormat:@"Connected: %@",
      peripheral.state == CBPeripheralStateConnected ? @"YES" : @"NO"];
    NSLog(@"%@", self.connected);
}
```

In this code, the `centralManager` calls `didConnectPeripheral` and instantiates it when a connection has been established. The delegate is set for the peripheral and noted in the log.

Looking Up Characteristics Associated with Services

Core Bluetooth creates a `CBService` object for each discovered service by calling `did DiscoverServices`. The following code discovers characteristics (see "Service and Characteristic Discovery" on page 69) and stores them in an array of `CBCharacteris tic` objects using the `discoverCharacteristics` method from the `CBPeripheral` class:

```
- (void)peripheral:(CBPeripheral *)peripheral
didDiscoverServices:(NSError *)error
{
    // sort through each service and discover chararacteristics
    // associated with each
    for (CBService *service in peripheral.services)
    {
        NSLog(@"Discovered service: %@", service.UUID);

        // discover characteristics associated with each service
        [peripheral discoverCharacteristics:nil forService:service];
    }
}
```

Use caution if you pass `nil` as a parameter to `[peripheral discoverServices:nil]` and `[peripheral discoverCharacteristics:nil forService:service]`, because the iOS device will then try to discover *all* the services and *all* the characteristics on the peripheral. This can waste battery power and time if the remote peripheral implements many more services and associated characteristics than the ones you're looking for. That's not a problem in this case, because we're implementing only the ones of interest on the peripheral side. But if it could be a problem in your particular application, use lists of services and characteristics instead.

Once the services of interest and the associated characteristics are discovered, you need to read the characteristic's values and make them available to the app. You can get the values directly using the `readValueForCharacteristic:` method, which is a good approach if the remote peripheral characteristic is static.

However, if the characteristic changes over time, as the battery level does, you should use the BLE notifications feature, described in "Server-Initiated Updates" on page 72. When this is done for the battery level characteristic, the application will be notified only when the value of the battery level changes. This avoids periodically polling the remote peripheral for changes, which can be hard on the battery life of the central iOS device because it creates unnecessary radio traffic.

You enable BLE notifications for a particular characteristic on the remote peripheral using the `setNotifyValue:forCharacteristic:` method (which writes to the corresponding CCCD on the peripheral, as explained in "Client Characteristic Configuration Descriptor" on page 62), as shown in the following code for the battery level characteristic (again, because the manufacturing data is static, it makes sense to use `readValueForCharacteristic:` for that):

```
- (void)peripheral:(CBPeripheral *)peripheral
didDiscoverCharacteristicsForService:(CBService *)service error:(NSError *)error
{
    // Retrieve Service for the Voltage Level
    if ([service.UUID isEqual:[CBUUID UUIDWithString:
                              BATTERY_LEVEL_SERVICE_UUID]])
    {
        for (CBCharacteristic *aChar in service.characteristics)
        {
            // Request level notifications
            if ([aChar.UUID isEqual:
                [CBUUID UUIDWithString:
                  BATTERY_LEVEL_MEASUREMENT_CHARACTERISTIC_UUID]])
            {
                [self.batteryPeripheral setNotifyValue:YES
                forCharacteristic:aChar];
                NSLog(@"Found battery level measurement characteristic");
            }
        }
    }


    if ([service.UUID isEqual:
        [CBUUID UUIDWithString:BATTERY_LEVEL_DEVICE_INFO_SERVICE_UUID]])
    {
        for (CBCharacteristic *aChar in service.characteristics)
        {
            if ([aChar.UUID isEqual:
                [CBUUID UUIDWithString:
                  BATTERY_LEVEL_MANUFACTURER_NAME_CHARACTERISTIC_UUID]])
            {
                [self.batteryPeripheral readValueForCharacteristic:aChar];
                NSLog(@"Found a device manufacturer name characteristic");
            }
        }
    }
}
```

Now, the `didUpdateValueForCharacteristic` method will be invoked whenever the peripheral has sent a notification, allowing the app to read the value of the battery level characteristic only when it changes:

```
- (void)peripheral:(CBPeripheral *)peripheral
didUpdateValueForCharacteristic:(CBCharacteristic *)characteristic
```

```
                              error:(NSError *)error
{
    // Updated value for battery level measurement received
    if ([characteristic.UUID isEqual:[CBUUID
        UUIDWithString:BATTERY_LEVEL_MEASUREMENT_CHARACTERISTIC_UUID]])
    {
        // Get the battery level
        [self getBatteryLevelData:characteristic error:error];
    }

}
```

To this point, the code has created services and characteristics objects associated with the remote peripheral. The next section uses those objects to supply data to the rest of the application.

Methods for Reading and Decoding Characteristics

This next section of code retreives the notified battery level value from the characteristic objects for use elsewhere in the app:

```
- (void) getBatteryLevelData:(CBCharacteristic *)characteristic
                       error:(NSError *)error
{
    // Get the Battery Level
    NSData *data = [characteristic value];
    const uint8_t *reportData = [data bytes];
    uint16_t level = 0;

    if ((reportData[0] & 0x01) == 0)
    {
        // Retrieve the level value for the Battery
        level = reportData[1];
    }
    else
    {
        level = CFSwapInt16LittleToHost(*(uint16_t *)(&reportData[1]));
    }

    // Display the battery level value to the UI if no error occurred
    if( (characteristic.value)  || !error )
    {
        self.batteryLevel = level;
        self.batteryLevel.text = [NSString stringWithFormat:@"%i %",
                                  batteryLevel];
    }
    return;
}
```

And this code retrieves the manufacturing data for use in the app:

```
- (void) getManufacturerName:(CBCharacteristic *)characteristic
{
    NSString *manufacturerName = [[NSString alloc]
                            initWithData:characteristic.value
                                    encoding:NSUTF8StringEncoding];
    self.manufacturer = [NSString stringWithFormat:@"Manufacturer: %@",
                        manufacturerName];
    return;
}
```

At this point, the iOS API has enabled the battery level voltage that was measured by the remote peripheral A/D converter to be transferred via BLE onto the iOS device, making it available to display as a graph, present as text, or store in a database.

iBeacon

iBeacon apps use the Core Location framework (*http://bit.ly/1giabJo*) features of iOS to provide navigation and location-based features to iOS and Android devices indoors, where access to cell tower signals and GPS might not work well or at all.

The iBeacon model offers a variety of new possibilities for connecting devices and communicating based on location, including interesting variants in retail settings. For example, permission-based marketing would allow retailers to push special offers and other information to customers (who have the corresponding app on their phones) based on their location in the store. In a museum, iBeacons could provide self-guided tours, delivering detailed information about nearby exhibits (in the form of text, audio, or video presentations) to visitors.

A device that implements iBeacon functionality broadcasts (see "Broadcast and Observation" on page 38) BLE advertising packets with the following four values included:

Proximity UUID
   A 128-bit value that uniquely identifies one or more beacons as a certain type or from a certain organization.

Major value
   An optional 16-bit unsigned integer that can group related beacons that have the same proximity UUID.

Minor value
   An optional 16-bit unsigned integer that differentiates beacons with the same proximity UUID and major value.

RSSI value
   Programmed into the beacon to facilitate determining distance from the beacon based on signal strength.

In the museum example, the proximity UUID would be associated with one particular museum, and major and minor numbers could be used to group or distinguish multiple beacons within the museum. In this case, the major value might specify a room in the museum, while the minor number might be associated with a particular exhibit in that room.

If mutiple beacons are installed this way, each associated with a particular room and exhibit, ranging performed by the iOS application on the receiving device (see "Ranging" on page 134) can distinguish between them, estimate the distance to each, and determine the location of the receiving device. One possible use of this infomation by the iBeacon app could be to deliver relevant information about the exhibit closest to the person holding the receiving iOS device.

Advertising

Typically, a beacon advertises only and doesn't provide other services or connectability (because BLE devices stop advertising once a connection is made). The advertising packet format shown in Figure 9-3 follows the standard BLE format for adverting packets and uses the Manufacturer Specific Data AD Type (see "Advertising Data Format" on page 48).

| iBeacon Advertising Packet Format | | | | BLUETOOTH SPECIFICATION V 4.0 [Vol 3] |
Byte (Octet)	Value	Meaning	Format	Appendix C (Normative): EIR and AD Formats
0	0x02	Advertising Field Length (2 bytes in this case)	Length = AD field type + AD Data	
1	0x01	Advertising Flags	AD field type	Table 18.1
2	0x06	LE General Discoverable Mode, BR/EDR Not Supported	AD Data	
3	0x1a	Advertising Field Length (26 bytes in this case)	Length = AD field type + AD Data	
4	0xff	Advertising Manufacture Specific Data	AD field type	Table 18.3
5	0x4c	Mfg Data	AD Data	
6	0x00	Mfg Data		
7	0x02	Mfg Data		
8	0x15	Mfg Data		
9	0x1a	iBeacon UUID		
10	0xcb	iBeacon UUID		
11	0xad	iBeacon UUID		
12	0x6e	iBeacon UUID		
13	0xe1	iBeacon UUID		
14	0xa5	iBeacon UUID		
15	0x48	iBeacon UUID		
16	0x38	iBeacon UUID		
17	0xa6	iBeacon UUID		
18	0x2a	iBeacon UUID		
19	0x22	iBeacon UUID		
20	0x03	iBeacon UUID		
21	0x5d	iBeacon UUID		
22	0x00	iBeacon UUID		
23	0xc3	iBeacon UUID		
24	0x5B	iBeacon UUID		
25	0x00	Major MSB		
26	0x03	Major LSB		
27	0x00	Minor MSB		
28	0x02	Minor LSB		
29	0xbx	RSSI (2's Complement of)		

Figure 9-3. The advertising packet format used by iBeacon

The typical iBeacon is a simple coin-cell-powered piece of BLE hardware (the BLE112 module, for example). It is also possible to program an iOS device to perform the iBeacon advertising service using the Core Bluetooth framework. This is useful for testing, but not very practical, because the iBeacon app must always be in the foreground.

If you do want to turn your iOS device into an iBeacon transmitter, you can use the Locate for iBeacon app (*http://bit.ly/1hq5BJS*) from Radius Networks. Another useful app from Radius Networks is MacBeacon for OS X, which allows your Mac to function as an iBeacon, saving your iOS device for testing your iBeacon app.

Ranging

Ranging is the process whereby the iBeacon app running on a receiving device (such as an iPhone) uses the strength of the received radio signal from a nearby beacon to estimate the distance between the receiving device and the beacon. The signal strength is measured in RSSI (received signal strength indication), a number in dBm that is available for every peripheral discovered by BLE apps running on iOS devices. In particular, the RSSI is tracked for every beacon in range of the iOS device running an iBeacon app. The measured RSSI from a particular beacon varies as the iPhone moves around a room. In general, the RSSI gets smaller as the distance between the iPhone and the iBeacon increases.

Beacons also broadcast an RSSI value in the advertising packet. The value of the RSSI number in this case is fixed and programmed into the beacon during manufacturing. The RSSI is determined by measuring the beacon's signal strength at a fixed distance of one meter, typically using an iPhone running special-purpose software. For example, the Locate for iBeacon app (*http://bit.ly/1hq5BJS*) from Radius Networks can be used for this.

The RSSI value is actually stored as a signed eight-bit integer representing dBm units in the advertising packet, but you don't have to know this or deal with the RSSI being a logarithmic scale and varying (generally) as the inverse square of the distance from the beacon to use it as those details are taken care of by iOS. The main purpose of this calibration is to take care of variation in signal output from beacon to beacon, due to individual differences in construction and performance of the radio chips.

The iBeacon app compares the measured RSSI to the expected value of the RSSI at one-meter broadcast in the advertising packet by the beacon to estimate the distance between the beacon and the iOS device. This method yields a fairly good estimation of the distance to a beacon (typically, within less than one meter) if all the beacons are calibrated.

The Core Location service provides classes and methods for ranging using iBeacons, as demonstrated in the following examples. You will want to use instances of the `CLBea conRegion` class and associated methods. With calibrated beacons and `LCBeaconRe gion`, you only have to think in terms of actual distance in meters and *beacon regions* (circular areas surrounding a particular iBeacon with a radius you set).

For example, you can have the iBeacon app trigger some action relative to an iBeacon region, such as an alert that tells you when you enter or leave a particular region with a desired radius from a particular beacon. Methods are also provided to tell the app which iBeacon is closest, again based on ranging values.

Implementing an iBeacon App

To implement the following iBeacon app, all you need is the Core Location framework, within which the key class references for the application include `CLLocationManager`, `CLBeaconRegion`, and `CLBeacon`. No parts of the Core Bluetooth framework are used directly. This iPhone app detects the presence of nearby beacons and determines which one is closest.

To test the app, we programmed several BLE112 modules (shown in Figure 9-4), powered by CR2032 coin cells for easy placement.

Figure 9-4. BLE112 modules programmed as iBeacons and powered by CR2032 coin cells.

 For testing, the Bluegiga BLE112 modules generated and transmitted the required iBeacon advertising packet. Complete code for both the iPhone app and the testing program for the BLE112 modules is provided in the GitHub repository for this book (*http://bit.ly/1qoj8Ed*).

First, you need to create and register the beacon region:

```
@implementation BobsBeaconTracker

- (instancetype)init
```

```
{
    self = [super init];
    if (self == nil) return nil;

    self.locationManager = [[CLLocationManager alloc] init];
    self.locationManager.delegate = self;

    self.beaconRegion = [[CLBeaconRegion alloc]
    initWithProximityUUID: [[NSUUID alloc]
                            initWithUUIDString:
                            @"E2C56DB5-DFFB-48D2-B060-D0F5A71096E0"]
                            identifier: @"Bobs Beacon default region"];

    self.beaconRegion.notifyEntryStateOnDisplay = YES;
    [self.locationManager startMonitoringForRegion:self.beaconRegion];

    return self;
}
```

Since each iBeacon app must use a particular proximity UUID that is hardcoded into the app through the initWithProximityUUID process earlier, it will respond only to beacons with that UUID (the UUID that is registered with iOS when the app is downloaded). This means that users have control, because they must download the app to use it (though, once installed, the app can still receive alerts even if it is not open or running). Without explicitly downloading the app, they will not be bombarded by unwanted iBeacon alerts and notices associated with other beacons with different proximity UUIDs.

As of iOS 7.1, the operating system itself registers the beacon regions that work with the app when the app is installed. Then, even if the app is suspended or not running, the system will wake up the iOS app to deal with entering or exiting a beacon region, normally within about 10 seconds.

In this next section of code, the CLBeacon class provides methods and objects to start and stop monitoring for particular iBeacons, as well as methods that determine which beacon is closest to the iOS device:

```
// start ranging beacons
- (void)locationManager:(CLLocationManager *)manager
        didEnterRegion:(CLRegion *)region
{
    if (![region.identifier isEqualToString:self.beaconRegion.identifier])
        return;

    [self.locationManager startRangingBeaconsInRegion:self.beaconRegion];

    NSLog(@"entered region");
}

// stop ranging beacons
```

```objc
- (void)locationManager:(CLLocationManager *)manager
        didExitRegion:(CLRegion *)region
{
    if (![region.identifier isEqualToString:self.beaconRegion.identifier])
    return;

    [self.locationManager stopRangingBeaconsInRegion:self.beaconRegion];

    NSLog(@"exited region");
}

// determine closest beacon

- (void)locationManager:(CLLocationManager *)manager
        didRangeBeacons:(NSArray *)beacons
                inRegion:(CLBeaconRegion *)region
{
    __block CLBeacon * closestBeacon ;

    if (beacons.count < 1) {
        closestBeacon = nil;
    }
    else
    {
        NSLog(@"locationManager didRangeBeacons: %@", beacons);

        [beacons enumerateObjectsUsingBlock:^(CLBeacon * beacon, NSUInteger idx,
                                     BOOL *stop)
        {
            if ((closestBeacon == nil) || (beacon.rssi > closestBeacon.rssi))
            {
                closestBeacon = beacon;
            }

            NSLog(@"closest beacon: %@", closestBeacon);
        }];
    }

    if (![self beacon:self.closestBeacon isSameAsBeacon:closestBeacon]) {
        self.closestBeacon = closestBeacon;
        [[NSNotificationCenter defaultCenter]
          postNotificationName:BobsBeaconTracker_ClosestBeaconChanged
          object:self];
    }
}

// compare beacons

- (BOOL)beacon:(CLBeacon *)beacon isSameAsBeacon:(CLBeacon *)otherBeacon
{
    return ([beacon.proximityUUID isEqual:otherBeacon.proximityUUID] &&
```

```
                (beacon.major == otherBeacon.major) &&
                (beacon.minor == otherBeacon.minor));
    }
```

If this code were embedded in a museum application, the app could now determine the closest beacons and then take the appropriate action (e.g., provide multimedia content that futher explains the exhibit).

Android devices can also make use of iBeacons, or BLE beacons in general. If you are interested in using iBeacons with Android, see the Radius Networks Android library of APIs to interact with iBeacons (*http://bit.ly/1jyl2PF*).

Apple Notification Center Service with an External Display

The Apple Notification Center Service (ANCS) function in iOS is the source of notifications displayed as a banner message along the top of the active screen (or in place of the entire active screen) for timely alerts (e.g., when you receive a text message, miss a call, or for a variety of other applications). For example, when you receive an incoming call, the ANCS temporarily replaces the active screen with the screen shown in Figure 9-5.

Figure 9-5. A notification of a phone call on an iPhone

When iOS 7 was introduced, Apple included a BLE interface to the ANCS (*http://bit.ly/1iBWqlG*) to route similar alerts to BLE connected accessories—for example, a BLE-enabled watch.

 In the ANCS context, the iOS device always acts as a GATT server, and the device that displays the notifications acts as a GATT client. For efficiency, describing how the ANCS works requires some terminology. For the purposes of this discussion, we will refer to the iOS device that sends the ANCS notifications as the *notification provider* (NP) and the accessory waiting to receive notifications (e.g., the BLE-enabled watch) as the *notification consumer* (NC).

Notifications displayed on the iOS device are also known as *iOS notifications*, whereas notifications sent via BLE GATT characteristic are known as *GATT notifications*.

Using ANCS does not require programming an app on the iPhone. Instead, the accessory advertises its request to receive notifications via an advertising packet, which includes the service UUID of the ANCS (7905F431-B5CE-4E99-A40F-4B1E122D00D0). Figure 9-6 shows the structure of the packet.

ANCS Notification Consumer Advertising Packet Format				BLUETOOTH SPECIFICATION Version 4.0 [Vol 3]
Byte (Octet)	Value	Meaning	Format	Appendix C (Normative): EIR and AD Formats
0	0x02	Advertising Field Length (2 bytes in this case)	Length = AD field type + AD Data	
1	0x01	Advertising Flags	AD field type	Table 18.1
2	0x02	LE General Discoverable Mode	AD Data	
3	0x05	Advertising Field Length (4 bytes in this case)	Length = AD field type + AD Data	
4	0x09	Advertising Complete Local Name	AD field type	Table 18.3
5	0x42	'B'	AD Data	
6	0x4f	'O' local name		
7	0x42	'B'		
8	0x11	Advertising Field Length (17 bytes in this case)	Length = AD field type + AD Data	
9	0x15	Advertising Service Solicitation	AD field type	Table 18.9
10	0xd0		AD Data	
11	0x00			
12	0x2d			
13	0x12			
14	0x1e			
15	0x4b			
16	0x0f	ANCS UUID - reverse order, LSB to MSB		
17	0xa4			
18	0x99			
19	0x4e			
20	0xce			
21	0xb5			
22	0x31			
23	0xf4			
24	0x05			
25	0x79			

Figure 9-6. The advertising packet format used by the NC

When the NP iOS device scans an advertising packet with the ANCS UUID from the NC, it connects to the NC accessory device. Now, a curious role reversal takes place (at least in the way we might normally think of the data flow between the central and peripheral). The NC (peripheral) becomes a GATT client, subscribing to services on the NP (ANCS services) and receiving notifications and data from the NP (central) iOS device. This is completely normal, because, as mentioned in "Roles" on page 51, the GAP and GATT roles are independent from each other.

On the NP (iOS central device), the Apple Notification Center Service UUID is 7905F431-B5CE-4E99-A40F-4B1E122D00D0, with the following associated characteristics/UUIDs:

Notification Source
> UUID 9FBF120D-6301-42D9-8C58-25E699A21DBD (notifiable)

Control Point
> UUID 69D1D8F3-45E1-49A8-9821-9BBDFDAAD9D9 (writeable with response)

Data Source
> UUID 22EAC6E9-24D6-4BB5-BE44-B36ACE7C7BFB (notifiable)

The Notification Source is a mandatory characteristic, while the others are optional. Normally, the accessory (the NC) will subscribe to the Service Changed characteristic of the GATT service to receive notifications of changes in the ANCS automatically from the Notification Source. Upon receiving notification of a new *GATT notifications* from ANCS, the NC can request more information. To find out more, the NC sends (writes) a message to the Control Point, including the notification ID of interest and a list of associated attributes the NC wishes to receive. Those are then provided in response from the Data Source.

The NC (peripheral) firmware must perform the following steps:

1. Start advertising (typically, once per second), including the ANCS UUID in the advertising packet to alert any NP in range.

2. When the NP (iOS device) connects, pair (if not already bonded) or enable encryption with the NC device.

3. Enumerate the iOS ANCS services.

4. Set the client for the notification source.

5. On notification, write a message to ANCS Control Point if more information is desired.

6. Receive response from the Data Source (containing notification ID and data).

Pairing of the NC (given local name BOB in this example) with the iOS device is straightforward. After the NC (remote peripheral) starts advertising, you must open Settings→Bluetooth on the NP (central) iPhone to trigger a scan for BLE devices and then manually pair the device (as shown in Figure 9-7). No PIN is required.

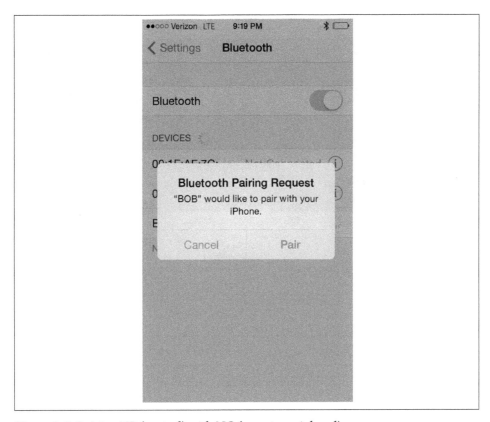

Figure 9-7. Pairing NP (central) with NC (remote peripheral)

Figure 9-8 shows the notification received on a terminal emulation screen connected to the NC (peripheral). In this case, the NP (an iPhone 5) sent three notifications: an incoming call, a missed a call, and a new voicemail message.

The first line, CALL (notification type), followed by the line ns, denotes the notification string received by the NC and written to the NC UART for display on the terminal emulator. The string done indicates that the connection for the notification is complete. Then the software scans the data sent along with the notification.

The next pieces of data are a UID (all zeros) and a text string with the name of the caller (caller ID): Davidson Ro. The next line, MISSED, is the beginning of a second notification event (for a missed call). The third notification (for a new voicemail message) begins with the line VMAIL.

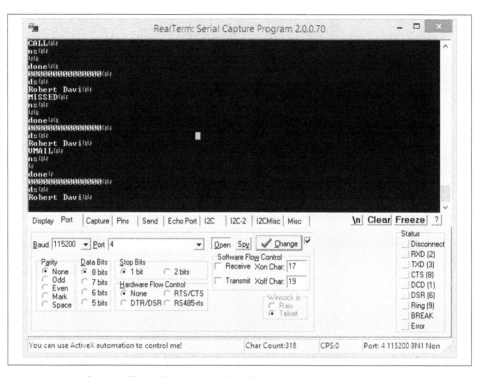

Figure 9-8. A phone call notification displayed on a terminal emulator screen

CHAPTER 10

Embedded Application Development

Focusing on readily available and open source embedded development kits and platforms, this chapter describes some of the tools available to anyone who wants to create custom firmware for Bluetooth Low Energy peripherals.

The first part of the chapter introduces a high-level BLE API that makes use of the mbed (*http://mbed.org*) development platform created and maintained by ARM. This is a great choice if you're new to embedded development, because it doesn't require familiarity with configuring an embedded toolchain yourself or working with embedded hardware on the lowest level. Most of the messy firmware implementation issues and setup problems are conveniently buried in the easy-to-use online tools and high-level APIs.

The second part describes embedded *toolchains*: collections of tools used together to convert standard source code into executable binaries that run on embedded processors. This section shows how to set up a cross-compiling toolchain to build ARM binaries on Windows, OS X, or Linux.

The last part of the chapter shows how to use these tools and concepts in the real world, taking advantage of a sample project for Nordic's nRF51822 system-on-chip (see "nRF51822-EK (Nordic Semiconductors)" on page 75) that allows you to transmit heart rate data to an iOS or Android device using the standard Heart Rate Profile (see "SIG-defined GATT-based profiles" on page 14 for more information on BLE profiles).

Complete code for the sample project is available in the GitHub repository for this book (*http://bit.ly/1qoj8Ed*).

mbed BLE API

As part of their objective to make developing embedded hardware using ARM Cortex processors as easy as possible, ARM and its partner companies created an open source development platform called mbed (*http://mbed.org*). mbed allows you to write code that is portable across a variety of supported ARM processors (*https://mbed.org/plat forms/*) and can take advantage of APIs and components built to run on top of these processors.

You can also use mbed with free online collaborative development tools and a variety of offline commercial and open source toolchains and IDEs. A considerable amount of effort has been put into defining high-level APIs that abstract away most of the low-level chip details that can consume a large part of the overall development budget. This enables reuse of open source software components that have been shared in the community and frees firmware engineers to focus more on project-specific code and less on low-level implementation details around their particular choice of microcontroller (MCU).

Most relevant to the purposes of this chapter, ARM recently added a BLE API to the mbed platform, which allows you to implement a simple GATT server with a few BLE services and characteristics (described in "Attribute and Data Hierarchy" on page 56) in a few dozen lines of code and a few hours of effort, without having to consider any vendor-specific details about the stack or chipset.

mbed provides an easy way to get a proof-of-concept product off the ground, while still working with a platform you can later move to production (exporting the code to offline compilers if necessary). The high-level abstractions around BLE mean you don't have to spend a lot of time learning about the specifics of your BLE SoC or module. For example, instantiating a standard BLE service or characteristic takes just one line of code:

```
GattService hrmService(GattService::UUID_HEART_RATE_SERVICE);
GattCharacteristic hrmRate(GattCharacteristic::UUID_HEART_RATE_MEASUREMENT_CHAR,
    2, 3, GattCharacteristic::BLE_GATT_CHAR_PROPERTIES_NOTIFY);
GattCharacteristic hrmLocation
    (GattCharacteristic::UUID_BODY_SENSOR_LOCATION_CHAR,
    1, 1, GattCharacteristic::BLE_GATT_CHAR_PROPERTIES_READ );
```

At the time of this writing, mbed's BLE API is still in beta and undergoing active development, but it currently covers most of the features you'll likely need for prototypes and proof-of-concept products. For more information on mbed or examples and updates on its BLE API, see the mbed project website (*http://mbed.org*).

Embedded Toolchains

While mbed qualifies as a viable embedded development platform with a clear path from prototype through to production, any vendor-neutral high-level API or platform necessarily requires some loss of control over the low-level implementation details. You generally trade ease of use for certain low-level optimizations in the code and drivers. This tradeoff makes sense in many situations, but obviously there is no one-size-fits-all approach with embedded development.

In the product design world, performance and control often trump ease of use, and many engineers still prefer to handle all of the implementation details themselves, writing their own low-level drivers and setting up their own build environment. Writing your own low-level drivers clearly requires more development effort, but it also ensures you have maximum control over the code running your products. It also forces you to understand the processor more intimately and allows you to optimize for performance and cost to a far greater extent.

Being able to fully optimize the code for size and performance is extremely important with embedded products. Optimizing for code size, for example, might allow you to use a smaller and cheaper processor, shaving $2 or $3 off the bill of materials, which can easily translate into $5 to $10 or more on the retail price. That savings can make all the difference between a product that succeeds or something that fails because it ends up in the wrong price bracket. The best way to control the final code size and performance is to have full control over your compiler and your build environment, allowing you to make the most out of every clock cycle on your embedded processor or SoC.

Compiling code for small embedded processors requires something called a *toolchain*. As the name implies, a toolchain is a collection of tools used to build executable code, and one of the most important parts of this toolchain is the *cross-compiler*. The cross-compiler compiles code while running on one architecture (for example, using an x86 instruction set) but produces code for a different architecture (for example, some variant of ARM). You have a number of commercial and open source options for cross compilers and low-level embedded toolchains, both commercial and open source, but we'll focus on an open source solution in this section.

In recent years, GCC (the free, open source compiler collection that's part of the GNU project) has made significant leaps and bounds in its support for ARM. Most of these advances are related to the dominance of ARM in the mobile phone and tablet arena (typically using ARM Cortex-A processors), but small, deeply embedded processors (ARM Cortex-M, etc.) have also benefited from the huge investments made here, due to overlapping instruction sets.

Widely used in a variety of industries, GCC is also conveniently available on any modern operating system and architecture. Code targetting GCC is generally highly portable, and you can build the same output on Linux, OS X, Windows, or almost any other environment you can imagine with no meaningful variation in the compiler output.

The latter point is extremely important and one of the main factors that makes GCC an excellent choice for embedded development. GCC does an excellent job with ARM processors today, but some commercial compilers still perform better at certain tasks. What GCC offers that no commercial toolchains can, though, is the assurance that you will *always* be able to rebuild your firmware using the same compiler version and dependencies. If commercial toolchains that require activation (and thus activation servers) aren't kept up to date, they might not be able to guarantee that they'll run on the current generation of operating systems or PCs.

If you're new to embedded development, it's easy to overlook this detail. Embedded devices can have lifespans of 10 or 20 years or more, far beyond the lifespan of most software packages. Commercial tools and development environments you use today might not exist in 10 years, and the vendor might not be around in the future to activate that long-retired product you invested heavily in years ago. GCC offers you the assurance that you'll never face this problem, since you can archive the full source of your cross-compiler, including any library dependencies, along with your firmware code and know you can rebuild it all at any point in the future.

Setting up a GNU toolchain on a non-Linux PC used to be a reasonably involved task, but ARM has made now made it trivial by providing precompiled, regularly updated versions of GCC for Windows, OS X, and Linux, including easy-to-use installers that take care of many of the messy details.

The first step in setting up a development environment for ARM is to download the latest prebuilt GNU toolchain (*https://launchpad.net/gcc-arm-embedded*). As shown in Figure 10-1, you can download convenient installers for OS X, Windows, and Linux, as well as the raw source code and a guide on building the toolchain yourself on other platforms.

The latest version available on the website might vary, but you should generally choose the most recent package. ARM provides quarterly updates, often incorporating improvements into the compiler and associated libraries, yielding smaller or more efficient compiled code.

Installing GNU Tools on OS X and Linux

If you're using OS X or Linux, you simply need to download the appropriate installer and run it. Any other development tools (`make`, various commands used in the makefile, etc.) are likely already available on your development machine or are easy to add (using Xcode on OS X and your package manager on Linux).

You can use the following command to confirm whether the GCC cross-compiler was successfully installed:

```
arm-none-eabi-gcc --version
```

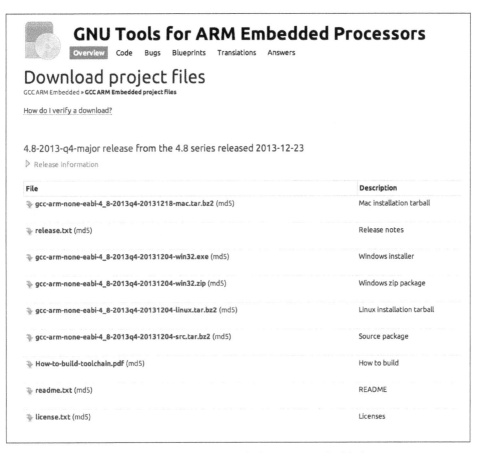

Figure 10-1. Download options for GNU Tools for ARM Embedded Processors

You should get a response that looks like this:

```
arm-none-eabi-gcc (GNU Tools for ARM Embedded Processors) 4.8.3 20131129
(release) [ARM/embedded-4_8-branch revision 205641]
Copyright (C) 2013 Free Software Foundation, Inc.
This is free software; see the source for copying conditions.  There is NO
warranty; not even for MERCHANTABILITY or FITNESS FOR A PARTICULAR PURPOSE.
```

If you see this output, it means that your cross-compiler has been sucessfully installed and you can generate ARM binaries on your development machine.

Installing GNU Tools on Windows

If you're using a Windows machine for development, you need to download the appropriate Windows installer and run it in the same manner you would any other installation package.

Make sure to select the post-installation option to "Add path to environment variable" (as shown in Figure 10-2). This will ensure that the toolchain is accessible from anywhere, which will generally make things easier when working with multiple folders and file locations.

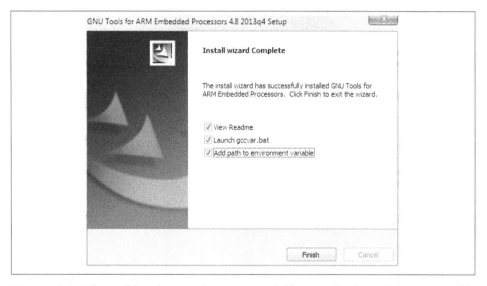

Figure 10-2. Select Add path to environment variable to make the toolchain accessible from anywhere

Similar to the OS X/Linux test shown previously, if everything is set up properly, you should be able to drop down to the command line and enter the following command to see which version of the GNU toolchain you've installed:

```
arm-none-eabi-gcc --version
```

Unlike Linux and OS X, Windows usually won't have some of the additional command-line tools the GNU toolchain generally requires, such as make and a number of *nix commands to work with files and directories (ls, cp, etc.). Thankfully, installing these additional tools on Windows is easy. GNU CoreUtils for Windows (*http://bit.ly/1ewD5kZ*) adds precompiled versions of all the file-editing commands you'll need.

Download the complete CoreUtils for Windows package, except sources, and run the installer.

You'll also need make, which you'll use to control the compilation process and convert your source code into something meaningful that you can run on the nRF51822. After installing GNU Coreutils for Windows, download and install Make for Windows (*http://bit.ly/Ra2fB2*).

You can test if make was installed succesfully by dropping down to the command line and entering the following command, which should give you a version number as a response:

```
make --version
```

Once both of these tools have been installed, you should have everything you need to cross-compile code for ARM on your development PC. Now, you can get started with the nRF51822 codebase and a sample project.

nRF51822 GNU Codebase and Sample Project

The ARM toolchain configured in the previous section allows you to build binaries for virtually any embedded device that uses an ARM processor. But to help give you an idea of what's involved with low-level embedded development in the real world, the following simple project is designed around Nordic's nRF58122-EK (see "nRF51822-EK (Nordic Semiconductors)" on page 75) using GCC and open source tools. The same principles and design process apply to any other embedded processor, though you will need to find startup code and tools from your silicon vendor or online to work with a different microcontroller or SoC.

The code base isn't exhaustive, but it should serve as a decent starting point for your own projects. It allows you to compile your firmware on a variety of platforms (Windows, OS X, or Linux) and provides a basic project structure that should be easy to understand and maintain.

 The nRF51822 GNU codebase, along with all the other sample code for this book, is hosted in the GitHub repository for this book (*http://bit.ly/1qoj8Ed*). It's worth looking at this repository for the latest code and any additional suggestions on working with the codebase that might have been added after publication of this book.

Going into exhaustive detail of Nordic's APIs and BLE stack is beyond the scope of this chapter and deserves a book in itself, but we've made a conscious effort to keep the sample code as minimal, clear, precise, and easy to understand as possible. The following sections cover how to set up your PC with everything you need to start working with this codebase, favoring easy-to-install tools to build a simple heart rate monitor project (*http://bit.ly/1kdO3kB*).

Getting the nRF51822 GNU Codebase

The nRF51822 GNU Codebase (*http://bit.ly/1kR2Pw9*) is available within the GitHub repository for this book (*http://bit.ly/1qoj8Ed*), along with all the other sample code from this book.

If you are using Linux or OS X, you probably already have Git available on the command line. To create a local copy of the repository that you can update as any changes are made to the codebase, simply execute the following command:

```
git clone git@github.com:microbuilder/IntroToBLE.git
```

To get the latest version of the code, go to the root folder of your project and enter the following command:

```
git pull
```

If you are on Windows, you can install Git from a precompiled binary (for example, msysgit (*http://msysgit.github.io/*)) and run the previous command.

If you prefer to avoid using Git and avoid version control altogether, you can also simply download an archive of the latest code by going to the GitHub repository (*http://bit.ly/1qoj8Ed*) and clicking Download ZIP.

nR51822 GNU Codebase Structure

Once you have a local copy of the nRF51822 codebase, you should end up with a file structure resembling Figure 10-3.

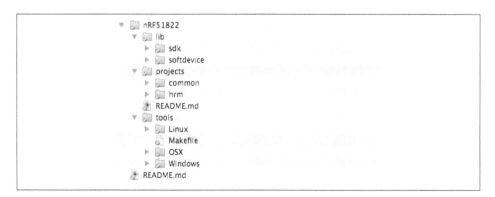

Figure 10-3. Bare file structure for the nRF51822 codebase

The *projects* folder contains sample projects. For any new project that you create based on this codebase, add a new folder here with a meaningful name describing your project. To help you get started, we've included a sample heart rate monitor project in the *hrm* folder. This project implements the standard heart rate service (*http://bit.ly/PXX0DD*) as defined by the Bluetooth SIG.

The *tools* folder gives you a convenient location to store OS-specific tools that you use during development. Including your tools here ensures that they are added to your version control system and guarantees they'll be available at any point in the future. You might even want to add the binaries for your GNU toolchain here, to avoid any ambiguity about the version used when building your firmware and make debugging easier months or years down the road.

The *lib* directory should contain the SDK, SoftDevice, and any other proprietary files from Nordic. You'll need to download these files from Nordic's website directly, because licensing issues prohibit us from including them.

 In order to download the SDK and an appropriate SoftDevice for the nRF51822 from Nordic Semiconductor's website (*http://www.nordicsemi.com/*) you will need to create a MyPages account and register the product key that was printed on the packaging for your nRF51822-EK. This product key gives you access to all of the resources available for this chipset, so be sure to keep this product key somewhere safe, in case you need to reuse it in the future.

After downloading the files, add them to the *lib* directory with the folder structure shown in Figure 10-4.

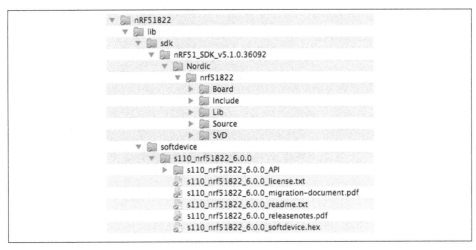

Figure 10-4. File structure for Nordic's SDK and SoftDevice

Once you've updated your project to include these files, you're ready to compile your first program.

Compiling Projects

If you have set up your GNU toolchain properly, including the ARM cross-compiler and any associated tools properly (as decribed earlier in this chapter), compiling should be as simple as dropping down to the command line, moving into the appropriate project folder (e.g., /projects/hrm), and running the following command:

```
make clean release
```

This will cause the make utility to interpret the *makefile*, which tells the compiler and toolchain exactly how to convert the source code in your folder into a binary image that can run on the target hardware. make starts the makefile interpreter, clean tells make to remove any previous build artifacts and start any cross-compilation with a clean slate, and release tells make to optimize the code for production situations, removing any unused code and extra debug information.

If this were a debug build, you could alternatively enter make clean debug, though this would produce larger executable code, because the binary data includes more debug information and unused code generally isn't removed.

If everything was set up correctly, your command-line results should look similar to Figure 10-5.

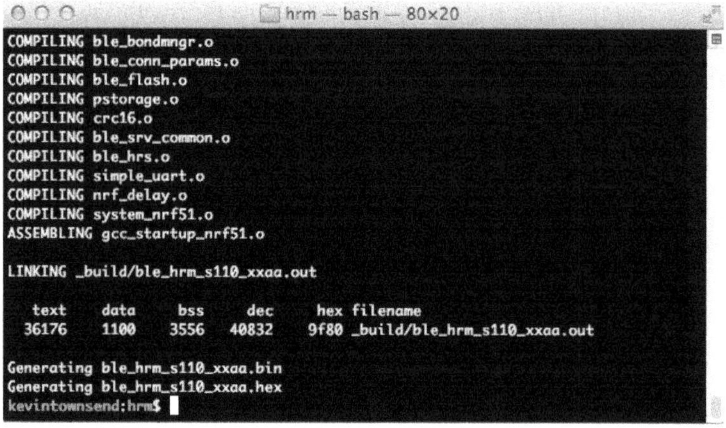

Figure 10-5. Successful cross-compilation and linking using GCC

This indicates that all of the .c files have been converted to object code (files with a .o extension), all of this object code has been assembled, and the linker has merged all of this data into a single file (*ble_hrm_s110_xxaa.out* in this case).

The following values are sizes (in bytes) that represent how much space the code took once compiled, assembled, and linked:

text

> The amount of data that ended up in flash memory, which consists of executable instructions and read-only data. Everything in this section will be written to flash memory.

data

> The amount of space used by initialized data, which are variables that have a specific value assigned to them at startup (for example, `int16_t i = 1023`, which is initialized to a specific value). Everything in this section will be stored in flash and then copied to SRAM, which is usually a precious resource on most small microcontrollers.

bss

> The amount of space used by uninitialized data, which are variables that don't have a value assigned to them (for example, `int16_t i`, which has no value assigned to it). Everything in this section will be allocated in SRAM.

dec

> The size of all data combined, including both flash and SRAM data.

The last two lines are the output from a helpful tool called `arm-none-eabi-objcopy`, which converts the *.out* file into other file formats that third-party tools might find easier to work with, including Intel Hex, which is a common file format when working with embedded systems.

Writing to the nRF51822

To write your program code to the non-volatile memory on the development board in the nRF51822-EK, you can use Nordic's nRFGo Studio utility, which is available on the same MyPages section of Nordic's website where you downloaded the SDK and Soft-Device (in "nR51822 GNU Codebase Structure" on page 150).

> nRFGoStudio is currently a Windows-only tool, but Roland King has created an OS X alternative called rknrfgo (*http://bit.ly/1qHOqUJ*). This unoffical application implements a subset of the functionality provided by nRFGo Studio, allowing you to program the flash memory on the nRF51822 using an easy-to-use GUI.

> You can also program the flash memory on the nRF51822 directly from the command-line on Windows, OS X, or Linux using Segger's J-Link drivers (*http://bit.ly/1ehaRjo*) and the associated tools. Documentation and examples of how to work with the J-Link are available on Segger's website (*http://bit.ly/1lVS3J4*).

If this is the first time that you are using your development board, you will need to write the SoftDevice *.hex* file to flash first. This writes Nordic's BLE stack to the lower half of flash memory on the device. To do this, find the *.hex* file for your SoftDevice (which was downloaded from Nordic's website as part of the SoftDevice package) and select it in the Program SoftDevice tab of nRFGoStudio, as shown in Figure 10-6.

Figure 10-6. Programming the SoftDevice onto the nRF51822 using nRFGo Studio

 The SoftDevice generally needs to be written only once, unless you want to update to a different version. The user code will need to be updated every time you make a change to your program, but this user code usually won't affect the SoftDevice, because they are stored in separate regions of flash.

Once the SoftDevice has been written, you can write your custom application code to the top half of memory, using the *.hex* file generated previously and the same nRFGo Studio tool, switching to the Program Application tab shown in Figure 10-7.

At this point, your application code has been written to the nRF51822 SoC and should start executing automatically. The application code will be able to make calls to the low-level SoftDevice for any BLE-specific functionality, and if everything is running properly, you should see LED0 on the PCA10001 development board blinking.

You can test the code you compiled by running an application on a BLE-enabled phone or tablet. First, download Nordic's nRF Utility from Apple's App Store (*http://bit.ly/1ewDMuD*) or Google Play (*http://bit.ly/PYggkr*). After installing the app, simply select HRM from the main menu and click Connect, which should show something similar to Figure 10-8.

Figure 10-7. Programming the application code onto the nRF51822 using nRFGo Studio

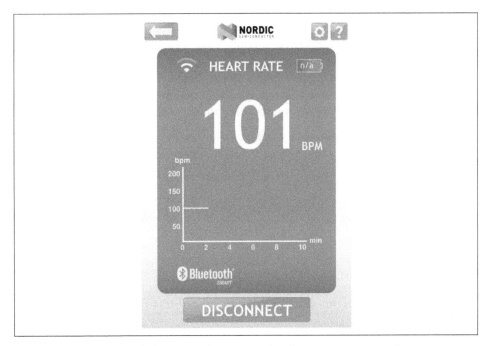

Figure 10-8. Using Nordic's nRF Utility to visualize heart rate monitor data

If you want to have a different look at exactly what your nRF51822 and application code are transmitting over the air, or if something doesn't look quite right to you, you can also try using any of the debug tools discussed in Chapter 6.

Going Further

Embedded development is a huge field, touching on a wide variety of domains. It involves physics with radio signal propogation and antenna design, electrical engineering with hardware design and part selection, mechanical engineering and industrial design with enclosures and overall product development, embedded software development for your firmware, manufacturing knowledge to source parts and assemble hardware, and effective test and validation strategies to make sure it doesn't all come back to you prematurely.

This tutorial covers only the tiniest part of embedded development, focusing on firmware design, but if you're interested in design and developing your own embedded hardware, there's probably never been a better time to get started, regardless of your technical background. Costs are coming down, information on both hardware and firmware development, as well as manufacturing know-how, have never been more accessible, and entire online ecosystems exist around these technologies.

If you're interested in going further with embedded development, take a look at some of the great open source hardware communities that have grown roots over the years, such as Adafruit (*http:://www.adafruit.com*) or Make (*http://makezine.com/*), as well as sites like Hackaday (*http://hackaday.com/*) that highlight news projects every day. Communities like this might get the ball rolling in your own head, or expose you to some ideas or technologies you might not have been familiar with before.

Nordic Semiconductor's also has a helpful Nordic Developer Zone (*http://bit.ly/1qomeI8*) forum, which can be a great source of answers to common problems working with this chipset.

Index

We'd like to hear your suggestions for improving our indexes. Send email to index@oreilly.com.

V

value (of a characteristic), 55
vendor-specific profiles, 14
vendor-specific UUIDs, 52

W

white list security feature, 23

Wibree, 1
Windows RT 8.1, Application Accelerator, 91
Wireshark, 87, 88, 93
write operations, 27

About the Authors

Kevin Townsend specializes in embedded design and development around the ARM Cortex-M family of microprocessors, and has a long-standing interest in low-power wireless communication. He's active in the open source hardware world as lead engineer at Adafruit Industries, where his job is taking interesting technologies in the embedded engineering space and getting them into the hands of domain experts in other fields to see what interesting solutions they can come up with when technology becomes invisible.

Carles Cufí has been involved with Bluetooth since 2000. Starting at Parrot in Paris with version 1.0 of the specification, he wrote one of the first protocol stacks to be shipped on a commercial product, and he has been involved with the development and implementation of Bluetooth devices and systems ever since. He is currently employed by Nordic Semiconductor, where he is responsible for the Bluetooth Low Energy Application Programming Interfaces offered to the developers using the nRF51 family of Integrated Circuits.

Akiba has been involved in wireless sensor networks since 2003. He wrote FreakZ, an open source Zigbee protocol stack, and also Chibi, an open source 802.15.4 protocol stack. He's a researcher for Keio University in the Internet and Society research group, and is a design consultant to the United Nations. His specialty and interest is in sensor networks for environmental monitoring. He currently runs FreakLabs, an open source wireless company, and is working on a hackerspace in the Japanese countryside called Hackerfarm.

Robert Davidson's passion in life is to apply what he knows about technology to solve real problems for people. He especially enjoys applications that use sensors to connect the physical world to computers and the Internet. He runs Ambient Sensors, a company focused on sensors and wireless sensor networks, and has a strong interest in the development of startup companies (and the scars to prove it). He enjoys sharing his interests and expertise with others.

Colophon

The animal on the cover of *Getting Started with Bluetooth Low Energy* is a mousebird, or Senegal coly (*Coliiformes*). This group of birds is found in sub-Saharan Africa, the only bird confined to that continent, though evidence shows they evolved in Europe.

The mousebird's feathers are described as "hair-like" and soft; they are slender, grey or brown, and soft, with crests and stubby bills. Typically growing to about 10 cm in length, it's characterized by its long, thin tail, which reaches another 20–24 cm past its body length. The mousebird feeds on berries, fruit, and buds, and scurry like rodents through leaves in search of food, which gave rise to the name mousebird. They are acrobatic, with strong claws and reversible out toes, and are able to feed upside down.

These birds are sociable, traveling in bands of around 20 in light woods. They inhabit grass-lined twig nests shaped like cups. They typically lay two to four eggs, and their young develop and acquire flight quickly.

The cover image is from Wood's *Animate Creation*. The cover fonts are URW Typewriter and Guardian Sans. The text font is Adobe Minion Pro; the heading font is Adobe Myriad Condensed; and the code font is Dalton Maag's Ubuntu Mono.

Get even more for your money.

Join the O'Reilly Community, and register the O'Reilly books you own. It's free, and you'll get:

- $4.99 ebook upgrade offer
- 40% upgrade offer on O'Reilly print books
- Membership discounts on books and events
- Free lifetime updates to ebooks and videos
- Multiple ebook formats, DRM FREE
- Participation in the O'Reilly community
- Newsletters
- Account management
- 100% Satisfaction Guarantee

Signing up is easy:

1. Go to: oreilly.com/go/register
2. Create an O'Reilly login.
3. Provide your address.
4. Register your books.

Note: English-language books only

To order books online:
oreilly.com/store

For questions about products or an order:
orders@oreilly.com

To sign up to get topic-specific email announcements and/or news about upcoming books, conferences, special offers, and new technologies:
elists@oreilly.com

For technical questions about book content:
booktech@oreilly.com

To submit new book proposals to our editors:
proposals@oreilly.com

O'Reilly books are available in multiple DRM-free ebook formats. For more information:
oreilly.com/ebooks

O'REILLY®

Lightning Source UK Ltd.
Milton Keynes UK
UKHW03f0300905 18
322282UK00005B/12/P